13th International Conference on Applied Physics and Mathematics (ICAPM 2023)

Journal of Physics: Conference Series Volume 2543

Singapore
10 - 12 March 2023

ISBN: 978-1-7138-7703-5
ISSN: 1742-6588

Printed from e-media with permission by:

Curran Associates, Inc.
57 Morehouse Lane
Red Hook, NY 12571

Some format issues inherent in the e-media version may also appear in this print version.

Institute of Physics
Dirac House, Temple Back
Bristol BS1 6BE UK

Phone: 44 1 17 929 7481
Fax: 44 1 17 920 0979

techtracking@iop.org

Additional copies of this publication are available from:

Curran Associates, Inc.
57 Morehouse Lane
Red Hook, NY 12571 USA
Phone: 845-758-0400
Fax: 845-758-2633
Email: curran@proceedings.com
Web: www.proceedings.com

TABLE OF CONTENTS

Preface

2023 13th International Conference on Applied Physics and Mathematics (ICAPM 2023) was held in Singapore during March 10-12, 2023. In the past years, ICAPM was held in many large capital cities, such as Tokyo, Phuket, Singapore, Busan, Lisbon, Hong Kong, and Bangkok, etc. This conference provides a remarkable opportunity for the academic and industrial communities to address new challenges and share solutions, and discuss future research directions.

The conference invited three internationally recognized experts in the field of applied physics and mathematics. They are Prof. Bakhodirzhon Siddikov, Ferris State University, USA, Prof. Naoyuki Ishimura, Chuo University, Japan, and Prof. Cheng-kuo Lee, National University of Singapore, Singapore. Scholars and researchers from China, Canada, India, Norway, Malaysia, Thailand, Indonesia, Philippines, Russia, and so on took this opportunity to share their research results and discuss the future development of related fields.

All papers in the conference proceeding have passed vigorous preliminary review and peer review by the technical committee. Professional comments are given according to the aspects of originality, innovation, applicability, technical merit, organizing and writing, relevance to conference, and so on. Authors benefit from the valuable comments and improve their submissions to meet the satisfaction of the reviewers. The proceeding contains a collection of research papers, and is divided into 3 chapters: Chapter 1 is entitled Basic Mathematics and Mathematical Computation; Chapter 2 is entitled Application of Mathematical Calculations and Models in Engineering; Chapter 3 is entitled Physical Theory and Engineering Physics.

We would like to especially thank the members of the Technical Committees. They worked very hard to review papers and made valuable suggestions for authors to improve their work. We would also like to thank the external reviewers for their assistance during the review process, and the authors for their contributions to the conference. Without such support, conferences cannot maintain quality or increase truly original contributions to knowledge. It's widely recognized that this is the main reason for its success, but we want to emphasize it again.

On behalf of organizing committee, we would like to express our sincere gratitude to all participants and speakers who contributed to make this event a great success. Looking forward to seeing you next year.

Po-Liang Liu
Conference Chair
ICAPM 2023

Journal of Physics: Conference Series **2543** (2023) 011001 doi:10.1088/1742-6596/2543/1/011001

Organizing Committees

Conference Chairs
Naoyuki Ishimura, Chuo University, Japan
Po-Liang Liu, National Chung-Hsing University, Taiwan

Program Chairs
Kuo-Chih Chuang, Zhejiang University, China
Jen Chuan Tung, Chang Gung University, Taiwan

TPC Members
Ilona Tregub, Financial University under the Government of the Russian Federation, Russia
Roberto Montemanni, University of Modena and Reggio Emilia, Italy
Xiaodi Li, Shandong Normal University, China
Rene Batac, De La Salle University, Philippines
Siti Suzilliana Putri Mohamed Isa, Universiti Putra Malaysia, Malaysia
Wael Zaki, Khalifa University, UAE
Darmesah Gabda, Universiti Malaysia Sabah, Malaysia
John Irish Lira, National University Manila Philippine, Philippines
Mazhar Hussain, Government College University, Pakistan
Zayid Abdulhadi, American University of Sharjah, UAE
Priscilla Macansantos, University of the Philippines Baguio, Philippines
Maria Cecilia Galvez, De La Salle University, Philippines
Zailan Siri, University of Malaya, Malaysia
Nik Mohd Asri Nik Long, Universiti Putra Malaysia, Malaysia
Dalal Maturi, King Abdulaziz University, Saudi Arabia
Reza Mohammadi, University of Neyshabur, Iran
Joginder Dhiman, Himachal Pradesh University, India
Ernest Macalalad, Mapúa University, Philippines
Johan Matheus Tuwankotta, Bandung Institute of Technology, Indonesia.
Haydar Akca, College of Arts and Sciences, Abu Dhabi University, UAE
Nor Haniza Sarmin, Universiti Teknologi Malaysia, Malaysia,
V. R. Lakshmi Gorty, SVKM's NMIMS University, India
Shijun Liao, Shanghai Jiao Tong University, China
Haitham Farok, Abu Dhabi University, UAE
Alexandra B. Santos-Putungan, University of the Philippines Los Banos, Philippines
Ibrahim Rose Irnawaty, Universiti Sains Islam Malaysia, Malaysia
Thana Yeeram, Mahasarakham University, Thailand
Fuaada Mohd Siam, Universiti Teknologi Malaysia, Malaysia
Natita Wangsoh, Pathumwan Institute of Technology, Thailand
Kwangwon Ahn, Yonsei University, South Korea
Wachirapond Permpoonsinsup, Pathumwan Institute of Technology, Thailand
Rubén Darío Santiago Acosta, Tecnológico de Monterrey, Mexico
Bushra Aliwi, University of Babylon, Iraq
Yohannes Gedamu, Bahir Dar University, Ethiopia

Journal of Physics: Conference Series **2543** (2023) 011001 doi:10.1088/1742-6596/2543/1/011001

Peer Review Statement

All papers published in this volume have been reviewed through processes administered by the Editors. Reviews were conducted by expert referees to the professional and scientific standards expected of a proceedings journal published by IOP Publishing.

- **Type of peer review:** Double Anonymous
- **Conference submission management system:** Morressier, Zmeeting
- **Number of submissions received:** 38
- **Number of submissions sent for review:** 38
- **Number of submissions accepted:** 20
- **Acceptance Rate (Submissions Accepted / Submissions Received × 100):** 52.6
- **Average number of reviews per paper:** 2
- **Total number of reviewers involved:** 56
- **Contact person for queries:**
  Name: Po-Liang Liu
  Email: pliu@dragon.nchu.edu.tw
  Affiliation: National Chung-Hsing University, Taiwan

The geometrical and Möbius-invariant properties of parabolic cycles

Sneha Gupta[1a*], Debapriya Biswas[1b]

[1]Department of Mathematics, Indian Institute of Technology Kharagpur, West Bengal-721302, India

Email: snehag863@gmail.com[a*], priya@maths.iitkgp.ac.in[b]

Abstract. We have studied parabolic cycles and given their equivalent matrix representations. Invariant properties of matrices under similarity have contributed to the Möbius-invariant properties in cycles. We have further discussed the inner product in the cycle space. Also, the geometrical properties of orthogonality and reflection have been studied to obtain the irregular points in the cycle space.

1. Introduction

The conic sections including straight lines are commonly represented by the term 'cycles', [1]. We have studied parabolic cycles corresponding to the system of dual numbers represented by the set $\mathbb{R}^p = \{u + \varepsilon v;\ u, v \in \mathbb{R}, \varepsilon^2 = 0\}$, where ε is called the dual unit. We have given an equivalent 2×2 matrix representation of a parabolic cycle to study the effect of the Möbius action, [2][3][4] of $SL(2;\mathbb{R})$, [5] on them. We have established Möbius action, [6] on cycles through similarity operation in matrices. Also, we have incorporated the similarity-invariant properties of a matrix like trace to define Möbius-invariant inner-product and consequently orthogonality in cycles. Lastly, we have defined reflection in cycles to obtain irregular points of the map connecting the cycle space to the cycle equation.

2. Parabolic cycles

A generic cycle is defined by the set of points $(u, v) \in \mathbb{R}^2$, explicitly given by the equation, [7]

$$k(u^2 - \sigma v^2) - 2lu - 2nv + m = 0;\ k, l, n, m \in \mathbb{R}\ (not all\ 0) \tag{1}$$

where $\sigma = -1, 0$ or 1. Parabolic cycles corresponding to dual numbers are defined by putting $\sigma = 0 = \varepsilon^2$ in the above equation i.e.,

$$ku^2 - 2lu - 2nv + m = 0;\ k, l, n, m \in \mathbb{R}\ (not all\ 0) \tag{2}$$

The parabolic cycle given by equation (2) can be equally well represented by the point $(k, l, n, m) \in \mathbb{R}^4$. For a non-zero scaling factor γ, the point $(\gamma k, \gamma l, \gamma n, \gamma m)$ defines an equivalent relation to equation (2). So, we consider the projective space $P(\mathbb{R}^4)$ i.e., the set of equivalence classes of $\mathbb{R}^4 \setminus \{0\}$ under the equivalence relation $(k, l, n, m) \sim (\gamma k, \gamma l, \gamma n, \gamma m)$ for any real $\gamma \neq 0$. In our terminology, the projective space $P(\mathbb{R}^4)$ is referred to as the cycle space and $\mathbb{R}^p$ as the point space. In order to perform various operations on cycles, we conveniently represent the quadruple (k, l, n, m) that defines a cycle, through the matrix $C_{\breve{\sigma}} = \begin{pmatrix} l+\breve{\varepsilon}n & -m \\ k & -l+\breve{\varepsilon}n \end{pmatrix}$ with a new dual unit $\breve{\varepsilon}$. $\breve{\sigma} = \breve{\varepsilon}^2$ has the values $-1, 0$ or 1 which are not dependent on those of ε^2. This matrix is the Fillmore-Springer-Cnops(FSC) matrix. We shall be using the following notations, [7]

(i) Q: The map which associates a point $(k, l, n, m) \in P(\mathbb{R}^4)$ i.e., the cycle space to the cycle equation

$$ku^2 - 2lu - 2nv + m = 0 \tag{3}$$

(ii) M: The map which associates a point $(k, l, n, m) \in P(\mathbb{R}^4)$ i.e., the cycle space to the matrix

$$C_{\breve{\sigma}} = \begin{pmatrix} l+\breve{\varepsilon}n & -m \\ k & -l+\breve{\varepsilon}n \end{pmatrix} \tag{4}$$

Correspondence between equation (3) and equation (4) i.e., between Q and M could be identified with Figure 1.

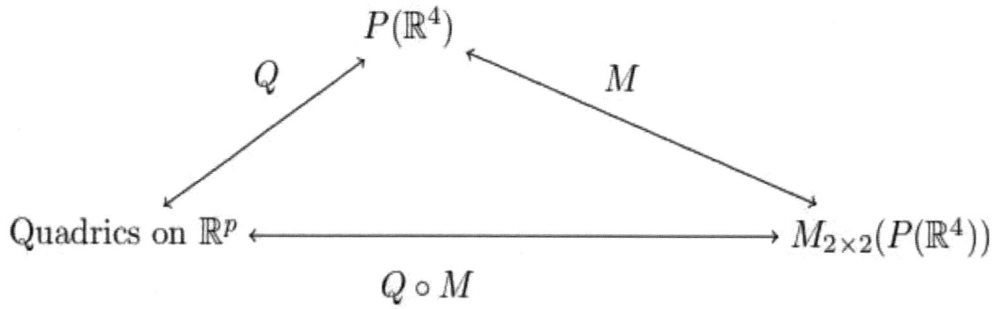

Figure 1. Correspondence between Q and M

*Remark 1.*The above correspondence enables us to write the same cycle in two equivalent forms i.e., $C_{\breve{\sigma}} = (k, l, n, m) \equiv \begin{pmatrix} l+\breve{\varepsilon}n & -m \\ k & -l+\breve{\varepsilon}n \end{pmatrix}$.

*Theorem 1.*The image $\tilde{C}_{\breve{\sigma}}$ of a cycle $C_{\breve{\sigma}}$ under $SL(2;\mathbb{R})$-action on $\mathbb{R}^p$ is given by the similarity

$$\tilde{C}_{\breve{\sigma}} = g C_{\breve{\sigma}} g^{-1} \tag{5}$$

In another words, Fillmore-Springer-Cnops construction explains the Möbius action of $SL(2;\mathbb{R})$ on cycles through the linear map given by equation (5) i.e., it intertwines them or connects them closely through a simple relation of similarity. We next discuss the invariant cycle product under Möbius transform along with its various associated features using the above theorem.

3. Invariant cycle product

The cycle $\breve{\boldsymbol{\sigma}}$-product of two parabolic cycles is given by, [7]

$$\langle \boldsymbol{C}_{\breve{\sigma}}, \tilde{\boldsymbol{C}}_{\breve{\sigma}} \rangle = -\boldsymbol{tr}(\boldsymbol{C}_{\breve{\sigma}} \overline{\tilde{\boldsymbol{C}}_{\breve{\sigma}}}) \tag{6}$$

i.e., the negative trace of the matrix product of $\boldsymbol{C}_{\breve{\sigma}}$ and the hypercomplex conjugation of $\tilde{\boldsymbol{C}}_{\breve{\sigma}}$. (The hypercomplex conjugation of a matrix $\boldsymbol{C}_{\breve{\sigma}}$ is obtained by performing individual hypercomplex conjugation $\overline{\boldsymbol{u} + \breve{\boldsymbol{\varepsilon}}\boldsymbol{v}} = \boldsymbol{u} - \breve{\boldsymbol{\varepsilon}}\boldsymbol{v}$ of all the entries of the matrix $\boldsymbol{C}_{\breve{\sigma}}$)

*Lemma 1.*The value of cycle product will remain the same if both matrices are replaced by their images under similarity with the same element $\boldsymbol{g} \in \boldsymbol{SL}(\boldsymbol{2};\mathbb{R})$, [7].

Proof. For an arbitrary $\boldsymbol{g} \in \boldsymbol{SL}(\mathbf{2};\mathbb{R})$ we need to prove that $\langle \boldsymbol{gC_{\breve\sigma}g^{-1}}, \boldsymbol{g\widetilde{C}_{\breve\sigma}g^{-1}}\rangle = \langle \boldsymbol{C_{\breve\sigma}}, \boldsymbol{\widetilde{C}_{\breve\sigma}}\rangle$.

$$\begin{aligned}\langle gC_{\breve\sigma}g^{-1}, g\widetilde{C}_{\breve\sigma}g^{-1}\rangle &= -tr\left(gC_{\breve\sigma}g^{-1}\overline{g\widetilde{C}_{\breve\sigma}g^{-1}}\right)\\ &= -tr(gC_{\breve\sigma}g^{-1}g\overline{\widetilde{C}_{\breve\sigma}}g^{-1})(\because g=\overline{g})\\ &= -tr(gC_{\breve\sigma}\overline{\widetilde{C}_{\breve\sigma}}g^{-1})\\ &= -tr(C_{\breve\sigma}\overline{\widetilde{C}_{\breve\sigma}})\ (\because \text{trace is invariant under similarity})\\ &= \langle C_{\breve\sigma}, \widetilde{C}_{\breve\sigma}\rangle.\end{aligned}$$

*Lemma 2.*The cycle $\breve{\boldsymbol\sigma}$-product of cycles defined by quadruples $(\boldsymbol{k},\boldsymbol{l},\boldsymbol{n},\boldsymbol{m})$ and $(\widetilde{\boldsymbol k},\widetilde{\boldsymbol l},\widetilde{\boldsymbol n},\widetilde{\boldsymbol m})$ is $\boldsymbol{-2l\tilde l + 2\breve\sigma n\tilde n + m\tilde k + k\tilde m}$.

Proof .Suppose, $\boldsymbol{C_{\breve\sigma}} = (\boldsymbol{k},\boldsymbol{l},\boldsymbol{n},\boldsymbol{m}) \equiv \begin{pmatrix} l+\breve\varepsilon n & -m\\ k & -l+\breve\varepsilon n\end{pmatrix}$ and
$\widetilde{C}_{\breve\sigma} = (\tilde k,\tilde l,\tilde n,\tilde m) \equiv \begin{pmatrix} \tilde l+\breve\varepsilon\tilde n & -\tilde m\\ \tilde k & -\tilde l+\breve\varepsilon\tilde n\end{pmatrix} \in M_{2\times2}(P(\mathbb{R}^4))$. Then,

$$\begin{aligned}\langle C_{\breve\sigma}, \widetilde{C}_{\breve\sigma}\rangle &= -tr(C_{\breve\sigma}\overline{\widetilde{C}_{\breve\sigma}})\\ &= -tr\begin{pmatrix} l+\breve\varepsilon n & -m\\ k & -l+\breve\varepsilon n\end{pmatrix}\begin{pmatrix} \tilde l-\breve\varepsilon\tilde n & -\tilde m\\ \tilde k & -\tilde l-\breve\varepsilon\tilde n\end{pmatrix}\\ &= -tr\begin{pmatrix} l\tilde l-\breve\varepsilon\tilde n l+\breve\varepsilon n\tilde l-\breve\sigma n\tilde n-m\tilde k & -l\tilde m-\breve\varepsilon n\tilde m+m\tilde l+\breve\varepsilon\tilde n\\ k\tilde l-\breve\varepsilon\tilde n k-l\tilde k+\breve\varepsilon n\tilde k & -k\tilde m+l\tilde l+\breve\varepsilon\tilde n l-\breve\varepsilon n\tilde l-\breve\sigma n\tilde n\end{pmatrix}\\ &= -2l\tilde l+2\breve\sigma n\tilde n+m\tilde k+k\tilde m.\end{aligned}$$

*Theorem 2.*The cycle product is a symmetric bilinear function of two cycles from $\boldsymbol{M_{2\times2}}(\boldsymbol{P}(\mathbb{R}^4)) \times \boldsymbol{M_{2\times2}}(\boldsymbol{P}(\mathbb{R}^4)) \to \mathbb{R}$, [7].

Proof. For parabolic cycles $\boldsymbol{C_{\breve\sigma}}, \boldsymbol{\widetilde{C}_{\breve\sigma}}, \boldsymbol{C^1_{\breve\sigma}}, \boldsymbol{C^2_{\breve\sigma}}$ and $\boldsymbol{C^3_{\breve\sigma}} \in \boldsymbol{M_{2\times2}}(\boldsymbol{P}(\mathbb{R}^4))$ and $\boldsymbol\lambda \in \mathbb{R}$ we have,

(a) $\langle C_{\breve\sigma}, \widetilde{C}_{\breve\sigma}\rangle = -tr\left(C_{\breve\sigma}\overline{\widetilde{C}_{\breve\sigma}}\right) = \overline{-tr\left(C_{\breve\sigma}\overline{\widetilde{C}_{\breve\sigma}}\right)}$ ($\because$ trace is a real number in our case)

$$\begin{aligned}&= -tr\overline{\left(C_{\breve\sigma}\overline{\widetilde{C}_{\breve\sigma}}\right)}\ (\because \overline{tr(M)} = tr(\overline{M}))\\ &= -tr\left(\overline{C_{\breve\sigma}}\widetilde{C}_{\breve\sigma}\right)\ (\because \overline{AB} = \overline{A}\,\overline{B} \text{ and } \overline{\overline{A}} = A)\\ &= -tr\left(\widetilde{C}_{\breve\sigma}\overline{C_{\breve\sigma}}\right)\ (\because tr(AB) = tr(BA))\\ &= \langle \widetilde{C}_{\breve\sigma},\ C_{\breve\sigma}\rangle.\end{aligned}$$

(b) $\langle (C^1_{\breve\sigma}+C^2_{\breve\sigma}), C^3_{\breve\sigma}\rangle = -tr((C^1_{\breve\sigma}+C^2_{\breve\sigma})\overline{C^3_{\breve\sigma}})$

$$\begin{aligned}&= -tr\left(C^1_{\breve\sigma}\overline{C^3_{\breve\sigma}} + C^2_{\breve\sigma}\overline{C^3_{\breve\sigma}}\right)\ (\because (A+B)C = AC+BC)\\ &= -tr\left(C^1_{\breve\sigma}\overline{C^3_{\breve\sigma}}\right) - tr\left(C^2_{\breve\sigma}\overline{C^3_{\breve\sigma}}\right)\ (\because tr(A+B) = tr(A)+tr(B))\\ &= \langle C^1_{\breve\sigma}, C^3_{\breve\sigma}\rangle + \langle C^2_{\breve\sigma}, C^3_{\breve\sigma}\rangle.\end{aligned}$$

(c) $\langle \lambda C_{\breve\sigma}, \widetilde{C}_{\breve\sigma}\rangle = -tr\left(\lambda C_{\breve\sigma}\overline{\widetilde{C}_{\breve\sigma}}\right) = -\lambda tr\left(C_{\breve\sigma}\overline{\widetilde{C}_{\breve\sigma}}\right) = \lambda\langle C_{\breve\sigma}, \widetilde{C}_{\breve\sigma}\rangle$. ($\because tr(\lambda A) = \lambda tr(A)$).

From (a), (b) and (c) we deduce that the cycle product is a symmetric bilinear function.

Lemma 3. $\langle \boldsymbol{C_{\breve\sigma}}, \boldsymbol{C_{\breve\sigma}}\rangle = \boldsymbol{2detC_{\breve\sigma}}$, [7].

Proof. $\boldsymbol{detC_{\breve\sigma}} = \boldsymbol{det}\begin{pmatrix} l+\breve\varepsilon n & -m\\ k & -l+\breve\varepsilon n\end{pmatrix} = \boldsymbol{-l^2+\breve\sigma n^2+mk}$.

Then, $\langle C_{\breve\sigma}, C_{\breve\sigma}\rangle = -tr(C_{\breve\sigma}\overline{C_{\breve\sigma}}) = -2l^2+2\breve\sigma n^2+2mk = 2\left(-l^2+\breve\sigma n^2+mk\right) = 2\ detC_{\breve\sigma}$.

Journal of Physics: Conference Series **2543** (2023) 012001 doi:10.1088/1742-6596/2543/1/012001

Corollary 1. For a parabolic cycle $C_{\breve{\sigma}}$, the inner product $\langle C_{\breve{\sigma}}, C_{\breve{\sigma}} \rangle$ is indefinite.
Proof. $\langle C_{\breve{\sigma}}, C_{\breve{\sigma}} \rangle = 2\, det C_{\breve{\sigma}} = 2\left(-l^2 + \breve{\sigma} n^2 + mk\right)$ which can be both < or > 0 depending on the values of $l, \breve{\sigma}, m$ or k.

Definition 1. (Zero-radius cycle) A cycle $Z_{\breve{\sigma}}$ is called a zero-radius cycle if the value of the cycle product of $Z_{\breve{\sigma}}$ with itself is zero.

Lemma 4. Zero-radius cycles form a Möbius-invariant family, [7].

Proof. Let a cycle $Z_{\breve{\sigma}}$ be a zero-radius cycle i.e., $\langle Z_{\breve{\sigma}}, Z_{\breve{\sigma}} \rangle = 0$ and $g \in SL(2;\mathbb{R})$. Then,

$$\langle g.Z_{\breve{\sigma}}, g.Z_{\breve{\sigma}} \rangle = \langle g Z_{\breve{\sigma}} g^{-1}, g Z_{\breve{\sigma}} g^{-1} \rangle \text{ (by Theorem 1)}$$
$$= -tr\left(g Z_{\breve{\sigma}} g^{-1} \overline{g Z_{\breve{\sigma}} g^{-1}}\right) = -tr\left(g Z_{\breve{\sigma}} \overline{Z_{\breve{\sigma}}} g^{-1}\right) = -tr(Z_{\breve{\sigma}} \overline{Z_{\breve{\sigma}}}) = \langle Z_{\breve{\sigma}}, Z_{\breve{\sigma}} \rangle = 0.$$

Hence, $\langle Z_{\breve{\sigma}}, Z_{\breve{\sigma}} \rangle = 0 \Rightarrow \langle g.Z_{\breve{\sigma}}, g.Z_{\breve{\sigma}} \rangle = 0$. Thus, they form a Möbius-invariant family.

Theorem 3. Parabolic zero-radius cycles are vertically upward parabolas that always touch the horizontal axis.

Proof. Parabolic zero-radius cycles $Z_{\breve{\sigma}} = \begin{pmatrix} l + \breve{\varepsilon} n & -m \\ k & -l + \breve{\varepsilon} n \end{pmatrix}$ are the ones obtained by putting $\breve{\sigma} = 0$ in the product ,$\langle Z_{\breve{\sigma}}, Z_{\breve{\sigma}} \rangle = 2\left(-l^2 + \breve{\sigma} n^2 + mk\right)$. This implies $-l^2 + mk = 0 \Rightarrow k = \frac{l^2}{m}$. Then, the cycle equation becomes

$$ku^2 - 2lu - 2nv + m = 0$$
$$\Rightarrow \frac{l^2}{m}u^2 - 2lu - 2nv + m = 0$$
$$\Rightarrow (lu)^2 + m^2 - 2lmu - 2nmv = 0$$
$$\Rightarrow (lu - m)^2 = 2nmv$$
$$\Rightarrow \left(u - \frac{m}{l}\right)^2 = 2\frac{nm}{l^2}v,$$

which are vertical parabolas having vertex $\left(\frac{m}{l}, 0\right)$ that are points on the horizontal axis i.e., parabolic zero-radius cycles always touch the horizontal axis (see Figure 2).

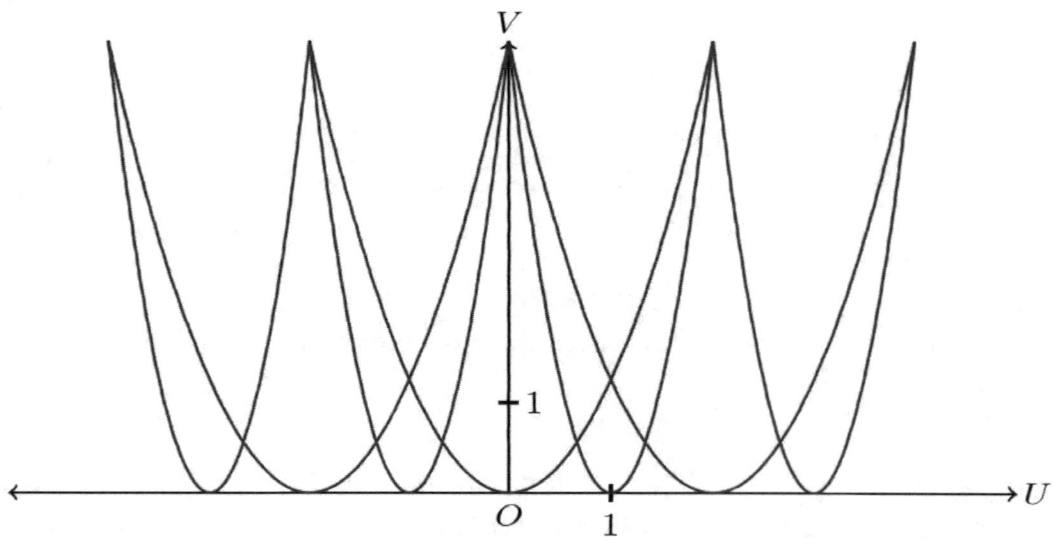

Figure 2. Parabolic zero-radius cycles

Since, we have discussed the cycle inner product, it comes to us naturally to discuss the cycle orthogonality as well. So, in the next section we define the orthogonality in cycles in terms of the inner product and show its various Möbius invariant properties.

4. $\breve{\sigma}$-orthogonality

Two cycles $C_{\breve{\sigma}}$ and $\widetilde{C}_{\breve{\sigma}}$ are $\breve{\sigma}$-orthogonal if their $\breve{\sigma}$-cycle product vanishes i.e., $\langle C_{\breve{\sigma}}, \widetilde{C}_{\breve{\sigma}} \rangle = 0$.

Lemma 5. The $\breve{\sigma}$-orthogonality condition is invariant under Möbius transformation, [7].

Proof. Let $C_{\breve{\sigma}}, \widetilde{C}_{\breve{\sigma}} \in M_{2\times 2}(P(\mathbb{R}^4))$ such that they are $\breve{\sigma}$-orthogonal i.e., $\langle C_{\breve{\sigma}}, \widetilde{C}_{\breve{\sigma}} \rangle = 0$. Then,
$\langle g.C_{\breve{\sigma}}, g.\widetilde{C}_{\breve{\sigma}} \rangle = \langle gC_{\breve{\sigma}}g^{-1}, g\widetilde{C}_{\breve{\sigma}}g^{-1} \rangle$ (by Theorem 1)

$$= -tr\left(gC_{\breve{\sigma}}g^{-1}\overline{g\widetilde{C}_{\breve{\sigma}}g^{-1}}\right) = -tr\left(gC_{\breve{\sigma}}\overline{\widetilde{C}_{\breve{\sigma}}}g^{-1}\right) = -tr\left(C_{\breve{\sigma}}\overline{\widetilde{C}_{\breve{\sigma}}}\right) = \langle C_{\breve{\sigma}}, \widetilde{C}_{\breve{\sigma}} \rangle = 0.$$

Hence, $\langle C_{\breve{\sigma}}, \widetilde{C}_{\breve{\sigma}} \rangle = 0 \Rightarrow \langle g.C_{\breve{\sigma}}, g.\widetilde{C}_{\breve{\sigma}} \rangle = 0$. Thus, they form a Möbius-invariant family.

Lemma 6. A cycle is $\breve{\sigma}$-self-orthogonal if and only if it is a zero-radius cycle, [7].

Proof. The proof is quite self-explanatory and trivial for if a cycle $C_{\breve{\sigma}}$ is self-orthogonal i.e.,$\langle C_{\breve{\sigma}}, C_{\breve{\sigma}} \rangle = 0 \Rightarrow C_{\breve{\sigma}}$ is a zero-radius cycle (by Definition 1). Conversely, let $C_{\breve{\sigma}}$ be a zero-radius cycle i.e., $\langle C_{\breve{\sigma}}, C_{\breve{\sigma}} \rangle = 0 \Rightarrow C_{\breve{\sigma}}$ is $\breve{\sigma}$-orthogonal to itself i.e., it is self-orthogonal.

If we consider the map Q, which connects the cycle space to the cycle equation then, we have an irregular point of this map given by the point $(0,0,0,1)$ for it corresponds to the equation $1 = 0$ which is absurd. We now try to obtain this irregular point geometrically by defining the reflection in zero-radius cycles.

5. Reflection in zero-radius cycles

Definition 2. (Reflection) The reflection of a zero-radius cycle $Z_{\breve{\sigma}}$ in a cycle $C_{\breve{\sigma}}$ is given by the similarity $C_{\breve{\sigma}}\overline{Z_{\breve{\sigma}}}C_{\breve{\sigma}}$, [7].

Definition 3. (Zero-radius cycle at infinity) The irregular point $(0,0,0,1) \in P(\mathbb{R}^4)$ of the map Q is called the zero-radius cycle at infinity and is denoted by Z_∞, [7].

Lemma 7. Z_∞ is the image of the zero-radius cycle $Z_{(0,0)} = (-1,0,0,0)$ at the origin under reflection into the unit cycle $(1,0,0,-1)$, cf.[7].

Proof. Given, $Z_{(0,0)} = (-1,0,0,0) \equiv \begin{pmatrix} 0 & 0 \\ -1 & 0 \end{pmatrix}$ (see Remark 1) and let $C_{\breve{\sigma}} = (1,0,0,-1) \equiv \begin{pmatrix} 0 & 1 \\ 1 & 0 \end{pmatrix}$.
Then, the reflection of $Z_{(0,0)}$ into the unit cycle $C_{\breve{\sigma}}$ is given by

$$C_{\breve{\sigma}}\overline{Z_{(0,0)}}C_{\breve{\sigma}} = \begin{pmatrix} 0 & 1 \\ 1 & 0 \end{pmatrix}\begin{pmatrix} 0 & 0 \\ -1 & 0 \end{pmatrix}\begin{pmatrix} 0 & 1 \\ 1 & 0 \end{pmatrix} = \begin{pmatrix} 0 & -1 \\ 0 & 0 \end{pmatrix} \equiv (0,0,0,1) = Z_\infty.$$

6. Conclusion and future work

We studied various $SL(2;\mathbb{R})$-invariant properties of parabolic cycles under Möbius action. The highlight of our work was the FSC construction which enabled us to perform various operations on cycles through matrices. Intertwining of Möbius action on cycles by similarity in matrices helped us to define various geometrical properties like orthogonality and reflection in cycles. The reflection property helped us in finding the irregular points in the cycle space with the help of the unit cycle. In future, we shall try to obtain more results related to reflection in cycles and use the irregular points to study compactification in the space of dual numbers.

References

[1] Kisil V V 2010 Erlangen program at large-1: Geometry of invariants (2005) *SIGMA Symmetry Integrability Geom. Methods Appl.* 6
[2] Dummit D S and Foote R M 2004 Abstract Algebra *John Wiley and Sons Inc.*
[3] Biswas D 2012 Projective coordinates and compactification in elliptic, parabolic and hyperbolic 2-D geometry *Journal of Applied Analysis* 18
[4] Kisil V V 2013 Induced Representations and Hypercomplex Numbers *Advances in Applied Clifford Algebras* 23 pp 417-440
[5] Lang S 1985 SL(2;R) Graduate text in mathematics *Springer New York* 105
[6] Ponnusamy S 2005 Foundations of Complex Analysis *Alpha Science International*
[7] Kisil V V 2012 Geometry of Möbius transformations: Elliptic, Parabolic and Hyperbolic Actions of SL(2;R) *Imperial College Press*

Singularity resolving in solution of the boundary integral equation in two-dimensional vortex methods

Yulia Izmailova*, Ilia Marchevsky and Kseniia Sokol

Bauman Moscow State Technical University, 2-nd Baumanskaya st., 5, 105005 Moscow, Russia

* Corresponding author's e-mail: yulia.izmailova@mail.ru

Abstract. The problem of 2D incompressible flow simulation around airfoils with sharp edges and corner point is considered. The solution of the boundary integral equation with respect to vortex sheet intensity arising in Lagrangian vortex method has weak singularity that cannot be resolved correctly in the framework of the existing Galerkin-type numerical schemes. It is shown that for piecewise-smooth bounded solutions the known schemes allow for solution reconstruction with high quality and provide the 2-nd order of accuracy, while for singular solution their order of accuracy goes down to the 1-st. A numerical scheme is suggested that allows for solution singularity resolving and provides the 2-nd order of accuracy. As a model problem, the added mass tensor components computation is considered, since its exact value is known for the Joukowsky wing airfoil with sharp edge (cusp point).

1. Introduction

Vortex particle methods of computational fluid mechanics [1] have rather narrow range of applicability: they are suitable for incompressible one-phase isothermal flows simulation. However, they are widely used in engineering applications [2, 3], especially connected with estimation of hydrodynamic loads acting on structures. Reviewing briefly their advantages, the following features should be pointed out:

- external and unbounded flows can be simulated with exact satisfaction of the perturbations decay boundary condition at infinity;
- efficient coupling schemes are developed that allow for solving fluid-structure interaction (FSI) problems with arbitrary translations, rotations, and deformations of the streamlined surface;
- the majority of computational resources is "concentrated" in the part of the flow domain with non-zero vorticity; such area is usually rather compact;
- vortex particle methods belong to a class of particle-based methods where Lagrangian particles are considered as vorticity carriers; it follows therefore that vortex methods provide rather small numerical diffusion.

Note that the most part of a wide list of researches that are devoted to vortex methods and which could have been cited here, is connected with the problems of simulation of vorticity field evolution in the flow domain. Their systematic review can be found in [4], and one can state that more or less efficient and accurate algorithms have been developed that allow for simulation of convective transfer and viscous diffusion of vorticity. At the same time, the problems connected to the streamlined body (airfoil) simulation are investigated much poorly. As a rule, in papers and monographs only some general words can be found, mainly pointing out that the problem of the airfoil simulation is equivalent to the solving of some boundary integral equation (BIE) with respect to double layer potential density, vortex sheet

intensity or vortex flux intensity. It is also usually mentioned that the BIE can be solved using similar approaches to those ones, which are applied in the framework of the boundary element method – BEM. At the same time, the results of the accuracy investigation of known numerical schemes (that can be applied in 2D problems being solved by vortex particle methods) performed for model problems with known exact solution [5], show that the error can be high enough, and it can be the most significant factor that bounds the accuracy of the flow simulation. A more detailed review of the problems being solved, as well as the VM2D code developed directly by the authors which allows simulating two-dimensional incompressible flows around movable and immovable airfoils, is presented in [6].
In the present paper, the original approach is suggested that allows for improving the accuracy of the BIE solution for airfoils with sharp edges or corner points where the exact solution has the singularity.

2. The governing equation

We consider an approach to incompressible flow simulation around an airfoil, according to which it is necessary to solve the BIE with respect to vortex sheet intensity. For simplicity, let us assume that the flow domain F is infinite and bounded only by the considered airfoil, and there is some known vorticity distribution $\Omega(\boldsymbol{\xi})$, $\boldsymbol{\xi} \in F$, in the flow. Then, the unknown vortex sheet intensity $\gamma(\mathbf{r})$ satisfies the vector boundary integral equation that arises from the no-slip condition on the airfoil surface line:

$$\oint_K \frac{\mathbf{k}\times(\mathbf{r}-\boldsymbol{\xi})}{2\pi|\mathbf{r}-\boldsymbol{\xi}|^2}\gamma(\boldsymbol{\xi})dl_\xi - \alpha(\mathbf{r})\gamma(\mathbf{r})\boldsymbol{\tau}(\mathbf{r}) = \mathbf{f}(\mathbf{r}), \quad \mathbf{r}\in K, \tag{1}$$

where K is the airfoil surface line; $\mathbf{k}$ is a unit vector orthogonal to the flow plane; $\alpha(\mathbf{r})$ is equal to the outer angle at the corresponding point divided by 2π (so, on smooth parts of the airfoil boundary $\alpha = 1/2$); $\boldsymbol{\tau}(\mathbf{r})$ is tangent unit vector defined such as $\mathbf{n}(\mathbf{r})\times\boldsymbol{\tau}(\mathbf{r}) = \mathbf{k}$; $\mathbf{n}(\mathbf{r})$ is outer normal unit vector; the right-hand side is expressed as

$$\mathbf{f}(\mathbf{r}) = \alpha(\mathbf{r})\mathbf{V}_K(\mathbf{r}) - \mathbf{V}(\mathbf{r}),$$

where $\mathbf{V}(\mathbf{r})$ is flow velocity at point $\mathbf{r}$, calculated according to the Biot – Savart law

$$\mathbf{V}(\mathbf{r}) = \mathbf{V}_\infty + \int_F \frac{\mathbf{k}\times(\mathbf{r}-\boldsymbol{\xi})}{2\pi|\mathbf{r}-\boldsymbol{\xi}|^2}\Omega(\boldsymbol{\xi})dS_\xi + \int_K \frac{\mathbf{k}\times(\mathbf{r}-\boldsymbol{\xi})}{2\pi|\mathbf{r}-\boldsymbol{\xi}|^2}\gamma^{att}(\boldsymbol{\xi})dl_\xi + \int_K \frac{\mathbf{r}-\boldsymbol{\xi}}{2\pi|\mathbf{r}-\boldsymbol{\xi}|^2}q^{att}(\boldsymbol{\xi})dl_\xi; \tag{2}$$

$\mathbf{V}_\infty$ is incident flow velocity; $\gamma^{att}(\boldsymbol{\xi})$ and $q^{att}(\boldsymbol{\xi})$ are the so-called attached vortex and source sheet intensities, which are tangent and normal components of the airfoil surface velocity $\mathbf{V}_K(\boldsymbol{\xi})$, respectively:

$$\gamma^{att}(\boldsymbol{\xi}) = \mathbf{V}_K(\boldsymbol{\xi})\cdot\boldsymbol{\tau}(\boldsymbol{\xi}), \qquad q^{att}(\boldsymbol{\xi}) = \mathbf{V}_K(\boldsymbol{\xi})\cdot\mathbf{n}(\boldsymbol{\xi}).$$

Note that in case of immovable airfoil the last three terms in the expression for the right-hand side are equal to zero; the assumption about the absence of the other surfaces in the flow is not essential; in order to take it into account, one should consider the system of the BIEs instead of one equation.
Traditionally, the following approach to numerical solving of the equation (1) is applied: firstly, it is projected onto the normal unit vector $\mathbf{n}(\mathbf{r})$, as the result, one obtains a singular BIE of the 1-st kind with the Hilbert-type kernel $P_n(\mathbf{r},\boldsymbol{\xi})$

$$\oint_K P_n(\mathbf{r},\boldsymbol{\xi})\gamma(\boldsymbol{\xi})dl_\xi = \mathbf{f}(\mathbf{r})\cdot\mathbf{n}(\mathbf{r}), \qquad P_n(\mathbf{r},\boldsymbol{\xi}) = -\frac{\boldsymbol{\tau}(\mathbf{r})\cdot(\mathbf{r}-\boldsymbol{\xi})}{2\pi|\mathbf{r}-\boldsymbol{\xi}|^2}, \qquad \mathbf{r},\boldsymbol{\xi}\in K. \tag{3}$$

For its solution special quadrature rules are used (usually, similar to central quadratures rule) that allows for estimating the Cauchy principal value of the corresponding integrals. The family of such numerical schemes we call hereinafter *N*-scheme. Detailed investigation of such schemes with the proof of their convergence can be found in [7]. Another approach [8] leads to the so-called *T*-schemes and starts with the projection of the equation (1) onto tangent unit vector $\boldsymbol{\tau}(\mathbf{r})$. This means that it is necessary to solve the BIE of the 2-nd kind with a kernel $P_\tau(\mathbf{r},\boldsymbol{\xi})$ that is bounded in the case of smooth airfoil or absolutely integrable in the case of airfoils with sharp edges or corner points:

$$\oint_K P_\tau(\mathbf{r},\boldsymbol{\xi})\gamma(\boldsymbol{\xi})dl_\xi - \alpha(\mathbf{r})\gamma(\mathbf{r}) = \mathbf{f}(\mathbf{r})\cdot\boldsymbol{\tau}(\mathbf{r}), \qquad P_\tau(\mathbf{r},\boldsymbol{\xi}) = \frac{\mathbf{n}(\mathbf{r})\cdot(\mathbf{r}-\boldsymbol{\xi})}{2\pi|\mathbf{r}-\boldsymbol{\xi}|^2}, \qquad \mathbf{r},\boldsymbol{\xi}\in K. \tag{4}$$

For such an equation the family of more accurate numerical schemes can be developed; they are mostly based on the Galerkin approach.

3. Smooth and non-smooth airfoils

Let us firstly touch the case of smooth airfoil. The solution of the BIE now is smooth enough, and it can be reconstructed in principally without significant issues. We note only that the implementation of the above mentioned *T*-schemes may be non-trivial and requires rather accurate integrals computation that arise in Galerkin-type procedure [9, 10]. Theoretic estimates, proven by numerical computations, show that the accuracy of numerical solution depends on the airfoil surface line discretization way. If it is replaced by a polygon consisting of straight panels, the order of accuracy (in L_1 norm) cannot be higher than the second: the first order of accuracy is achieved for piecewise-constant solution representation, the second order of accuracy — for piecewise-linear distribution of the vortex sheet intensity [10]. In much more complicated schemes where the curvilinearity of the panels is taken into account explicitly [11], the third order of accuracy can be achieved at piecewise-quadratic solution representation.

So, one can conclude that the problem of higher-order numerical schemes development is solved, more or less, for smooth airfoils.

At the same time, when one deals with the airfoil with sharp edge (*e.g.*, Joukowsky wing airfoil) or corner points (*e.g.*, rectangular or semicircular airfoil), the situation becomes more dramatic.

Let us firstly examine the case of Joukowsky wing airfoil, or some similar one. Initially, we note that the BIE (1), as well as its projections onto both normal and tangent directions (3) and (4), has infinite set of solutions. In order to pick out the unique solution having physical sense, it is necessary, as a rule, to specify the total value of vorticity that is contained in the vortex sheet,

$$\oint_K \gamma(\xi)dl_\xi = \Gamma, \tag{5}$$

or, the same, to specify the velocity field circulation around the airfoil.

It is well-known that for wing airfoil there exists the unique value of circulation Γ^* that corresponds to the Chaplygin — Joukowsky condition, known also as the Kutta condition, according to which the flow velocity remains finite at the sharp edge [12] (note that this condition is also applicable to a wider class of the airfoils with single corner points, e.g., the so-called generalized Joukowsky airfoils [13]). In this case, the solution of the BIE (1) remains bounded and can has only jump discontinuity at the mentioned point, figure 1. For such solutions the above mentioned numerical schemes remain suitable, and they provide (sometimes with some obvious limitations) the same order of accuracy as for smooth solutions.

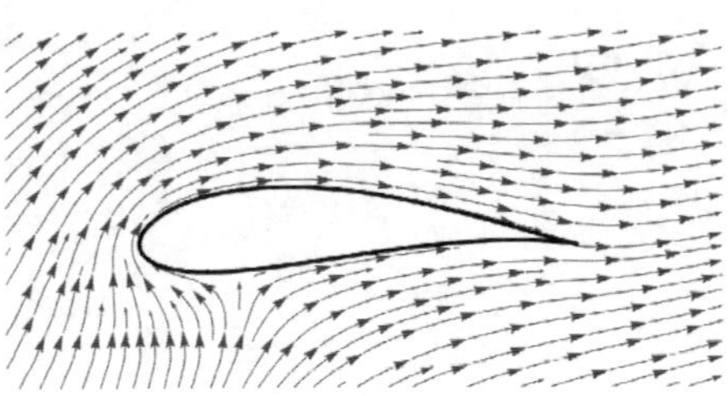

a

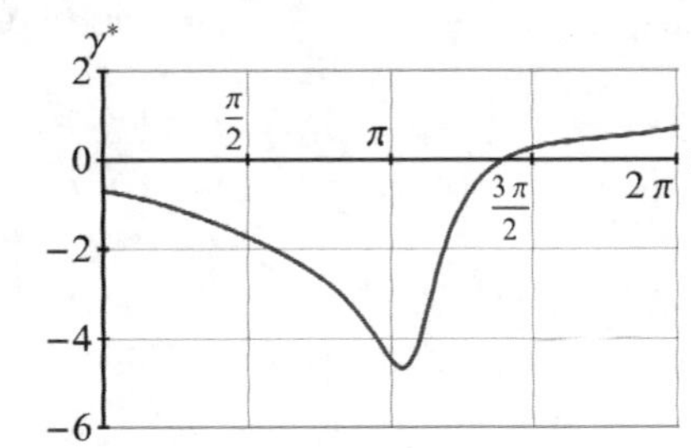

b

Figure 1. Flow around the Joukowsky wing airfoil satisfying the Chaplygin — Joukowsky condition (*a*) and vortex sheet intensity (*b*) against a parameter that varies from 0 to 2π along the airfoil surface line.

However, the Chaplygin — Joukowsky condition corresponds to the steady-state flow regime that, of course, can be interesting in some engineering problems, but such approach is not applicable to unsteady flows simulation, for example, when it is necessary to simulate transient regimes, *etc*. In this case, the value of circulation Γ depends only on angular acceleration of the airfoil (*i.e.*, if it moves plane-parallel or with constant angular velocity, then $\Gamma = 0$).

Thereby, for unsteady flows around the airfoils with edges or corner points the solution of the BIE (1) becomes singular at the corresponding points, figure 2, and its asymptotic behavior is known [7]:

$$\gamma(s) \sim s^{-\mu}, \quad \mu = 1 - \pi/\chi, \tag{6}$$

where s is the distance to the edge; χ is the outer angle at the edge point. Consequently, for the classical Joukowsky airfoil with sharp edge (that geometrically corresponds to cusp point with zero inner angle and outer angle $\chi = 2\pi$) there will be singularity of $s^{-1/2}$ type; for corner points with non-zero inner angle it has type $s^{-\mu}$, $0 < \mu < 1/2$.

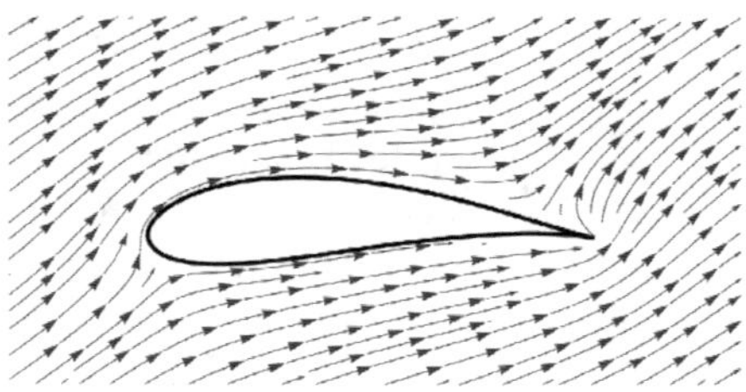

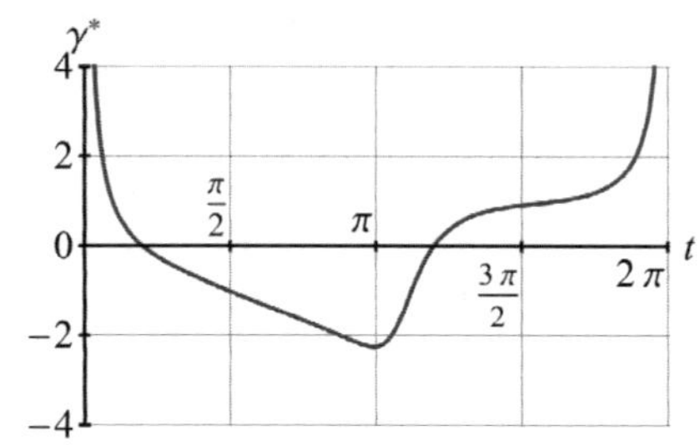

Figure 2. Flow around the Joukowsky wing for $\Gamma = 0$ (a) and vortex sheet intensity (b) against a parameter that varies from 0 to 2π along the airfoil surface line.

4. Galerkin-type numerical schemes

In the above mentioned numerical schemes the Galerkin approach is used to derive the linear system that is the discrete analogue of the BIE. The airfoil surface line K is replaced with the polygon consists of the panels K_i, $i = 1,\dots,N$; the solution is reconstructed as a combination of basis functions:

- for piecewise-constant solution representation

$$\gamma(\mathbf{r}) = \sum_{i=1}^{N} \gamma_i^0 \phi_i^0(\mathbf{r}), \quad \mathbf{r} \in K, \tag{7}$$

- for piecewise-linear solution representation

$$\gamma(\mathbf{r}) = \sum_{i=1}^{N} \left(\gamma_i^0 \phi_i^0(\mathbf{r}) + \gamma_i^1 \phi_i^1(\mathbf{r}) \right), \quad \mathbf{r} \in K, \tag{8}$$

where two families of basis functions are the following:

$$\phi_i^0(\mathbf{r}) = \begin{cases} 1, & \mathbf{r} \in K_i, \\ 0, & \mathbf{r} \notin K_i; \end{cases} \qquad \phi_i^1(\mathbf{r}) = \begin{cases} \dfrac{(\mathbf{r} - \mathbf{c}_i) \cdot \tau_i}{L_i}, & \mathbf{r} \in K_i, \\ 0, & \mathbf{r} \notin K_i, \end{cases}$$

here $\mathbf{c}_i$ is the center of the panel K_i; L_i is its length.

In order to determine values of coefficients γ_i^p, the solution in the form (7) or (8) is substituted to the T-model (4), then the expression for the residual is multiplied by a projection function $\psi_i^p(\mathbf{r})$, and the result is assumed to be equal to zero. As the result, the algebraic linear system is obtained, from which the unknown coefficients are found.

The most obvious way is to choose the projection functions equal to basis ones; this corresponds to the classical Galerkin approach and for smooth airfoils (and, consequently, smooth and bounded solutions) can be considered as residual orthogonality condition in a subspace of the L_2 functional space.

Such approach is implemented in [9, 10] where all necessary formulae for the integrals arising in the Galerkin procedure are presented. It works perfectly, providing the first and second orders of accuracy in case of smooth or at least bounded solutions in L_1 norm.

As an example, let us consider the problem of the added masses calculation. The added masses are widely used in various engineering applications in order to estimate the hydrodynamic forces acting the

structure that moves in the flow with acceleration, *i.e.*, vibrating in the flow. At the same time, according to [14] the added mass tensor components also can be considered as integral characteristics of the vortex sheet intensity on the airfoil surface line that satisfies the BIE (1) or equivalent equations (3) and (4).
In order to estimate the added masses of the airfoil, three problems should be considered: impulsive start of the airfoil in horizontal and vertical directions in still media with unit velocity $|\mathbf{V}_K|=1$, and its impulsive start in rotational motion with unit angular velocity $\omega=1$. The value of total circulation in additional condition (5) should be chosen as

$$\Gamma = -\oint_K \gamma^{att}(\xi)dl_\xi;$$

thus, it is equal to zero except of rotational motion.
Then, the components of the added mass tensor for two-dimensional flow can be calculated [14]:

$$\lambda_{dx} = \oint_K \rho y\big(\gamma_d(\mathbf{r}) + \gamma_d^{att}(\mathbf{r})\big)dl_r, \qquad \lambda_{dy} = -\oint_K \rho x\big(\gamma_d(\mathbf{r}) + \gamma_d^{att}(\mathbf{r})\big)dl_r,$$

$$\lambda_{d\omega} = -\frac{1}{2}\oint_K \rho(x^2+y^2)\big(\gamma_d(\mathbf{r}) + \gamma_d^{att}(\mathbf{r})\big)dl_r,$$

where x and y are the abscise and the ordinate of the point $\mathbf{r}$ at the airfoil boundary; index d has values x, y and ω and corresponds to the motion direction of the airfoil; $\gamma_d(\mathbf{r})$ and $\gamma_d^{att}(\mathbf{r})$ denote vortex sheets intensities in the airfoil motion in the *d*-th direction; ρ is the flow density.
The exact values are known for the added masses of the elliptic airfoil with semiaxes a and b [15]:

$$\lambda_{xx}^* = \rho\pi b^2, \quad \lambda_{yy}^* = \rho\pi a^2, \quad \lambda_{\omega\omega}^* = \frac{1}{8}\rho\pi(a^2-b^2)^2, \quad \lambda_{xy}^* = \lambda_{x\omega}^* = \lambda_{y\omega}^* = 0.$$

In the Table 1 relative error of the added masses computation is shown (maximal value for all diagonal components) for the elliptic airfoil with $5:1$ semiaxes ratio which surface line is discretized into N panels of equal length. $\mathcal{T}^0$ and $\mathcal{T}^1$ denote numerical schemes with piecewise-constant and piecewise-linear solution representations. In brackets a'posteriori estimations for the order of accuracy are shown.

Table 1. Relative errors of the added mass tensor estimation for the elliptical airfoil with 5:1 semiaxes ratio for different number of panels *N*.

N	100	200	400	800	1600	3200
$\mathcal{T}^0$	$1.69\cdot10^{-2}$	$4.39\cdot10^{-3}$	$1.10\cdot10^{-3}$	$2.73\cdot10^{-4}$	$6.81\cdot10^{-5}$	$1.70\cdot10^{-5}$
	–	(1.94)	(2.00)	(2.01)	(2.01)	(2.00)
$\mathcal{T}^1$	$9.16\cdot10^{-3}$	$2.52\cdot10^{-3}$	$6.48\cdot10^{-4}$	$1.64\cdot10^{-4}$	$4.11\cdot10^{-5}$	$1.03\cdot10^{-5}$
	–	(1.86)	(1.96)	(1.98)	(1.99)	(2.00)

One can see that the 2-nd order of accuracy takes place for both numerical schemes; the scheme $\mathcal{T}^1$ is $40\,\%$ more accurate in comparison to $\mathcal{T}^0$. Note that the order of accuracy for the $\mathcal{T}^0$ scheme is higher than for vortex sheet intensity; this is due to the fact that the solution is smooth. For the $\mathcal{T}^1$ scheme the orders of accuracy are equal to 2 in both cases.
Now let us examine these schemes for Joukowsky wing airfoil [15], which shape is shown in figure 3.

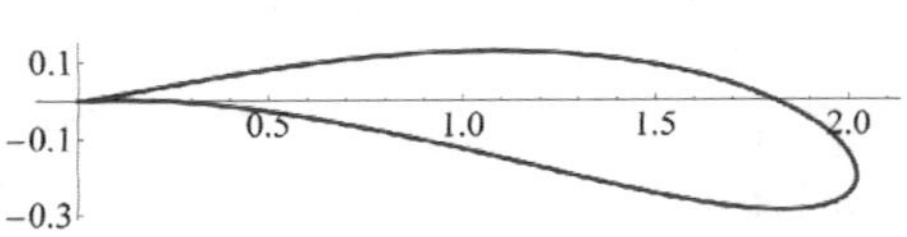

Figure 3. The Joukowsky airfoil shape for parameters $a=1$, $\eta=1.15$, $\alpha=\pi/30$.

The specific feature of this airfoil is that in horizontal motion at $\Gamma=0$ in condition (5) the exact solution for vortex sheet intensity $\gamma(s)$, where s is the arclength coordinate, is bounded and has jump discontinuity at the sharp edge (cusp point), figure 4, *a*.

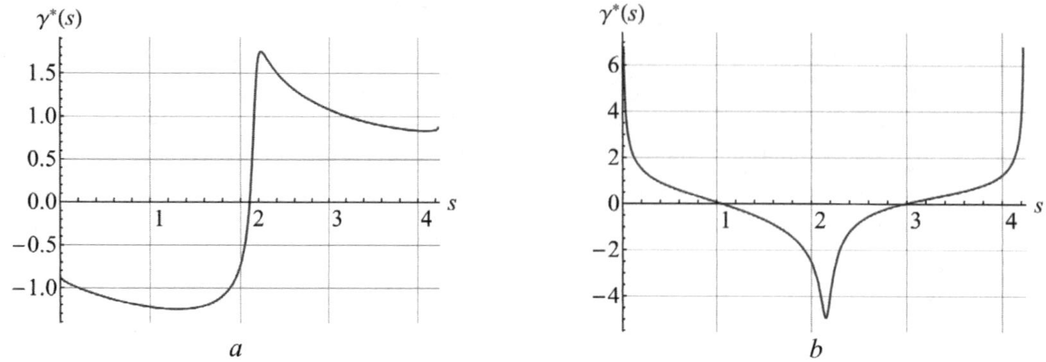

Figure 4. Exact solution for the vortex sheet intensity $\gamma^*(s)$ for the Joukowsky wing airfoil in horizontal (*a*) and vertical (*b*) motions; s is the arc length measured along the airfoil surface line from the sharp edge.

The expressions for added mass tensor components for such airfoil could be found in [15].
In the Table 2 maximal relative error of the added masses computation is given for λ_{xx}, λ_{xy} and $\lambda_{x\omega}$ (the components that can be obtained after horizontal airfoil motion simulation) for the wing airfoil discretized into N panels of equal length. Again, in brackets a'posteriori estimations for the order of accuracy are shown.

Table 2. Relative errors of the added masses λ_{xx}, λ_{xy} and $\lambda_{x\omega}$ estimation for the Joukowsky wing airfoil for different number of panels N.

N	100	200	400	800	1600	3200
T^0	$6.77\cdot10^{-3}$	$1.61\cdot10^{-3}$	$3.71\cdot10^{-4}$	$8.25\cdot10^{-5}$	$1.92\cdot10^{-5}$	$4.79\cdot10^{-6}$
	–	(2.07)	(2.12)	(2.17)	(2.10)	(2.00)
T^1	$4.55\cdot10^{-3}$	$1.18\cdot10^{-3}$	$2.99\cdot10^{-4}$	$7.50\cdot10^{-5}$	$1.88\cdot10^{-5}$	$4.70\cdot10^{-6}$
	–	(1.95)	(1.98)	(1.99)	(2.00)	(2.00)

Here we see that both schemes again provide the 2-nd order of accuracy and their errors are nearly the same for fine discretization.
However, if we consider all the components of the added mass tensor, which calculation requires vertical and rotational airfoil motion simulation, where solution is unbounded near the sharp edge (figure 4, *b*), we obtain much coarser result, Table 3.

Table 3. Maximal relative errors for all the added masses estimation for the Joukowsky wing airfoil for different number of panels N.

N	100	200	400	800	1600	3200
T^0	$2.51\cdot10^{-2}$	$1.29\cdot10^{-2}$	$6.88\cdot10^{-3}$	$3.73\cdot10^{-3}$	$2.03\cdot10^{-3}$	$1.11\cdot10^{-3}$
	–	(0.96)	(0.91)	(0.88)	(0.87)	(0.87)
T^1	$4.55\cdot10^{-3}$	$1.65\cdot10^{-3}$	$7.40\cdot10^{-4}$	$3.51\cdot10^{-4}$	$1.71\cdot10^{-4}$	$8.49\cdot10^{-5}$
	–	(1.46)	(1.16)	(1.08)	(1.03)	(1.01)

The scheme T^1 is now approximately 10 times more accurate than T^0, but both schemes provide now not higher than the 1-st order of accuracy due to not enough accuracy of solution reconstruction on the panels that are adjacent to the sharp edge. The numeriacal solution behaviour near the edge point at mesh refinement is shown in figure 5.

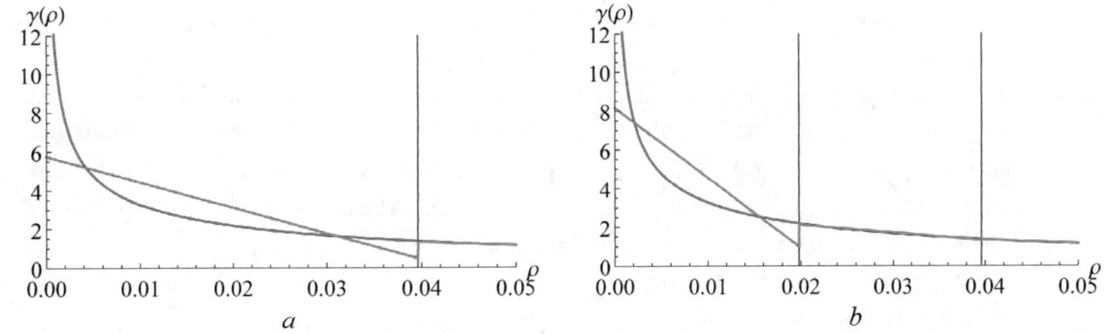

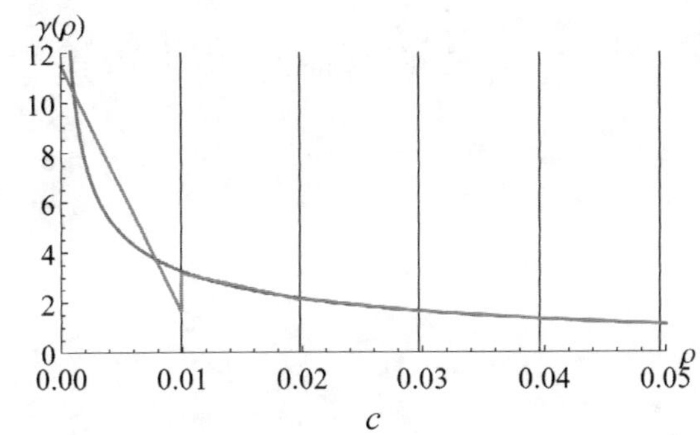

Figure 5. Vortex sheet intensity against the distance to the sharp edge obtained by using the $\mathcal{T}^1$ scheme with piecewise-linear solution representation (green line) and the exact solution (red curve); number of panels $N = 100$ (a), $N = 200$ (b), and $N = 400$ (c); vertical lines correspond to panel endings.

Such an issue cannot be overcome in the framework of the schemes $\mathcal{T}^0$ and $\mathcal{T}^1$, since the solution, that is unbounded now, belongs to L_1 functional space instead of space containing bounded smooth functions, or piecewise-smooth bounded functions with jump discontinuities at specified point.

5. Numerical scheme with singular basis functions

It is seen from figure 5, that the scheme $\mathcal{T}^1$ is quite suitable over all the panels except those, which are adjacent to the sharp edge. This put us onto an idea of the modification of this scheme that allows for taking into account singularity of the solution. Let us suggest the following representation for the solution:

$$\gamma(\mathbf{r}) = (\gamma_1^0\phi_1^0(\mathbf{r}) + \gamma_1^a\phi_1^a(\mathbf{r})) + \sum_{i=2}^{N-1}(\gamma_i^0\phi_i^0(\mathbf{r}) + \gamma_i^1\phi_i^1(\mathbf{r})) + (\gamma_N^0\phi_N^0(\mathbf{r}) + \gamma_N^a\overline{\phi}_N^a(\mathbf{r})), \qquad \mathbf{r} \in K, \tag{9}$$

where on the 1-st and the N-th panels the unbounded basis functions

$$\phi_1^a(\mathbf{r}) = \left(\frac{L_1}{s_1(\mathbf{r})}\right)^{\mu} - \frac{1}{1-\mu}, \qquad \overline{\phi}_N^a(\mathbf{r}) = \left(\frac{L_N}{L_N - s_N(\mathbf{r})}\right)^{\mu} - \frac{1}{1-\mu}$$

are introduced instead of linear ones; $s_i(\mathbf{r})$ means the distance from the beginning of the i-th panel to the point $\mathbf{r}$. Such choice of basis functions is consistent with the asymptotic behavior of the exact solution (6); the constant $(1-\mu)^{-1}$ is added in order to provide their average value equal to zero.

However, the traditional Galerkin approach, according to which projection functions are equal to basis ones, now does not seem to be suitable. First of all, we note that it is applicable in principle only for $\mu < 1/2$; this means that we can consider only airfoils with corner points, but not with sharp edges (cusp points), since in the last case the integrals over the 1-st and the N-th panels

$$\int_{K_1} (\phi_1^a(\mathbf{r}))^2\, dl_r = \infty, \qquad \int_{K_N} (\overline{\phi}_N^a(\mathbf{r}))^2\, dl_r = \infty,$$

Journal of Physics: Conference Series **2543** (2023) 012002 doi:10.1088/1742-6596/2543/1/012002

i.e., they do not converge. This follows from the fact that the exact solution for the airfoil with sharp edge ($\mu = 1/2$) does not belong to L_2 functional space, but belongs only to L_1 space.

In order to develop universal algorithm, we use the Petrov — Galerkin approach, according to which the set of projection functions differs from the set of basis ones. Namely, we use now the same projection functions as in original $\mathcal{T}^1$ scheme: $\{\phi_i^0\}_{i=1}^N$ and $\{\phi_i^1\}_{i=1}^N$, *i.e.*, piecewise-constant and piecewise-linear ones. Another advantage of such approach is connected to the computation of the coefficients of the resulting matrix: only $4N$ coefficients should be replaced by new values, while $4N(N-1)$ remain the same as in the matrix arising in the $\mathcal{T}^1$ scheme (note that the right-hand side remains exactly the same). The results of the BIE solution obtained by using the suggested scheme (let us denote it $\mathcal{T}_a^1$) for the above considered wing airfoil with cusp point are shown in figure 6.

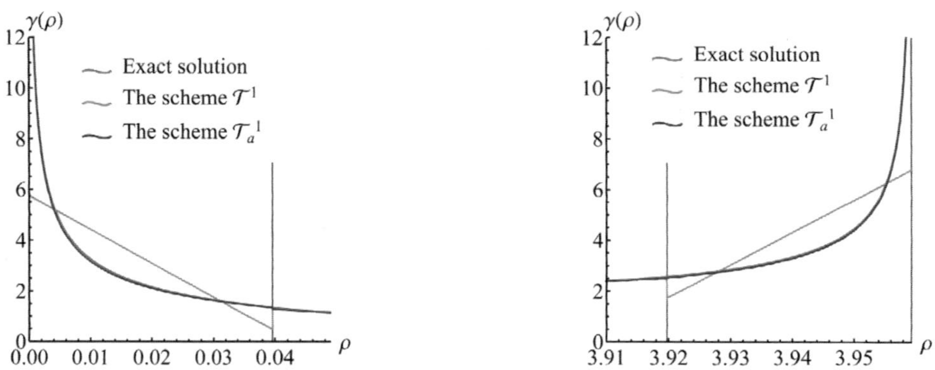

a *b*

Figure 6. Vortex sheet intensity over the 1-st (*a*) and the *N*-th (*b*) panels of the wing airfoil for the surface line discretized into $N = 100$ panels; vertical lines correspond to the panel endings.

It is clearly seen that the quality and accuracy of solution reconstruction on the 1-st panel is now much higher than for original $\mathcal{T}^1$ scheme. Over the *N*-th panel the exact and numerical solutions behavior is quite similar.

The importance of correct resolving of unbounded vortex sheet intensity over panels adjusting to the shapr edge becomes clear from figure 7 where the streamlines are reconstructed in neighborhood of such point according to the Biot – Savart law (2).

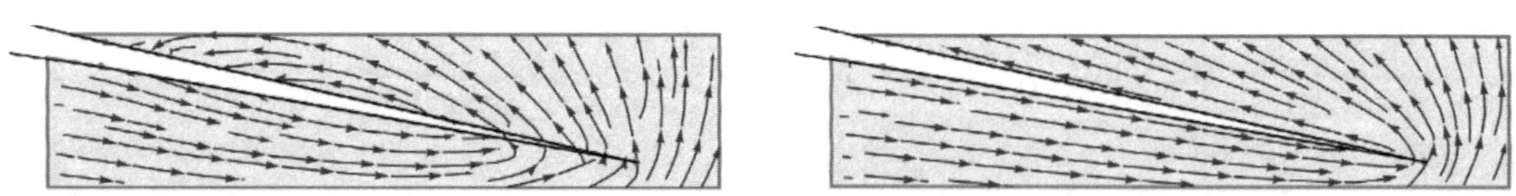

a *b*

Figure 7. The streamlines of velocity field for piecewise-linear (*a*) BIE solution reconstruction and with correct resolution of singularity (*b*) by using the developed scheme.

It is seen that the velocity field corresponding to piecewise-linear solution of the BIE (figure 7, *a*) is physically implausible since some “vortex” is generated and the no-through (no-penetration) boundary condition is not satisfied. The suggested scheme provides much more accurate result (figure 7, *b*).

Finally, let us now estimate the quality of the developed scheme $\mathcal{T}_a^1$ by estimating the added mass components tensor for the Joukowsky airfoil with sharp edge that was considered in the previous section (figure 3). The maximal relative error for all the added masses are shown in Table 4. It is seen that exactly the 2-nd order of accuracy is provided due to solution singularity accounting.

Table 4. Maximal relative errors of the all the added masses estimation for the Joukowsky wing airfoil for different number of panels.

N	100	200	400	800	1600	3200
$\mathcal{T}_a^1$	$4.59\cdot10^{-3}$	$1.19\cdot10^{-3}$	$3.01\cdot10^{-4}$	$7.56\cdot10^{-5}$	$1.89\cdot10^{-5}$	$4.74\cdot10^{-6}$
	–	(1.95)	(1.98)	(1.99)	(2.00)	(2.00)

6. Conclusion

The problem of the numerical solution of a boundary integral equation of the second kind arising in 2D vortex methods with respect to the intensity of the vortex sheet at the airfoil surface line is considered. In case of the airfoil with a corner point or a sharp edge, the exact solution of the integral equation becomes unbounded that does not allow the boundary integral equation to be numerically solved with high accuracy using the Galerkin approach with bounded basis functions. A scheme based on the Petrov — Galerkin method is proposed as an original numerical scheme that allows one to reconstruct unbounded solutions. In this scheme, constant basis functions and unbounded basis functions with a known asymptotic behavior are introduced on the panels where the solution is unbounded, on the other panels constant and linear basis functions are used; on all panels, constant and linear functions are chosen as projection functions. The developed scheme allows one to improve the accuracy significantly.

References

[1] Cottet G-H and Koumoutsakos P 2000 *Vortex methods: theory and practice* (Cambridge: CUP)
[2] Lewis R I 1991 *Vortex element methods for fluid dynamic analysis of engineering systems* (Cambridge: CUP)
[3] Branlard E 2017 *Wind turbine aerodynamics and vorticity-based methods: Fundamentals and recent applications* (Berlin: Springer)
[4] Mimeau C and Mortazavi I 2021 A review of vortex methods and their applications: from creation to recent advances *Fluids* **6** 68 doi 10.3390/fluids6020068
[5] Kuzmina K, Marchevsky I and Ryatina E 2019 Exact solutions of boundary integral equation arising in vortex methods for incompressible flow simulation around elliptical and Zhukovsky airfoils *J. Phys.: Conf. Ser.* **1348** 012099 doi 10.1088/1742-6596/1348/1/012099
[6] Marchevsky I, Sokol K, Ryatina E, Izmailova Y 2023 The VM2D open source code for two-dimensional incompressible flow simulation by using fully Lagrangian vortex particle methods *Axioms* **12** 248 doi 10.3390/axioms12030248
[7] Lifanov I K 1996 *Singular integral equations and discrete vortices* (Utrecht: VSP)
[8] Kempka S N et al. 1996 Accuracy considerations for implementing velocity boundary conditions in vorticity formulations *SANDIA report* **0583** p 52 doi 10.2172/242701
[9] Kuzmina K S and Marchevskii I K 2019 On the calculation of the vortex sheet and point vortices effects at approximate solution of the boundary integral equation in 2D vortex methods of computational hydrodynamics *Fluid Dyn.* **54** pp 991–1001 doi 10.1134/S0015462819070103
[10] Marchevskii I K, Sokol K S and Izmailova Yu A 2022 T-schemes for mathematical modelling of vorticity generation on smooths airfoils in vortex particle methods *Herald of the Bauman Moscow State Tech. Univ. Ser. Natural Sci.* **6** pp 33–59 doi 10.18698/1812-3368-2022-6-33-59
[11] Marchevsky I K, Kuzmina K S and Soldatova I A 2018 Improved algorithm of boundary integral equation approximation in 2D vortex method for flow simulation around curvilinear airfoil *Math. and Math. Modeling* **6** pp 22–51 doi 10.24108/mathm.0618.0000158
[12] Tokaty G A 1994 *A history and philosophy of fluid mechanics* (New-York: Dover pub.)
[13] Glauert H 1924 A generalised type of Joukowski aerofoil *NACA Reports and Memoranda* **911**
[14] Dynnikova G Y 2019 Added mass in a model of a viscous incompressible fluid *Doklady Physics* **64** pp 397–400 doi 10.1134/S1028335819100045
[15] Sedov L I (1965) *Two-dimensional problems in hydrodynamics and aerodynamics* (Geneva: John Wiley & Sons Inc.)

Hybrid Barnes – Hut/Multipole algorithm application to vortex particles velocities calculation and integral equation solution

Alexandra Kolganova*, Ilia Marchevsky and Evgeniya Ryatina

Bauman Moscow State Technical University, 2-nd Baumanskaya st., 5, 105005 Moscow, Russia

* Corresponding author's e-mail: kolganova.ao@mail.ru

Abstract. The approximate fast algorithm is developed that makes it possible to calculate velocities of the vortex particles in two-dimensional flow simulations by using vortex methods. It is adapted to efficient solution of the boundary integral equation arising at each time step of simulation. The proposed method can be considered as generalization of the classical Barnes – Hut method, taking into account some ideas of the Fast Multipole Method. Some model problems are considered that can be solved by using vortex methods, the developed fast algorithm is implemented for multi-cores CPUs and (with some modifications) for GPUs. The scalability of the code is rather high; the developed algorithms can run with millions of vortex particles taking only tens or hundreds of milliseconds per time step; linear systems with dimension of about 10^4 can be solved on multicore CPU in tens milliseconds.

1. Introduction

Vortex particle method [1-3] seems to be an efficient tool for hydro- and aerodynamical loads estimation that act on structures in flows. Such problems are especially complicated in coupled statement, when the law of motion for the structure (or some element of structure) is unknown and it is determined by these loads. Existing "engineering" approaches, such as preliminary knows stationary hydrodynamic coefficients usage, are applicable only for steady-state regimes simulation, and do not allow to estimate the structure behavior at transient regimes; it is also hardly possible to simulate correctly, even for steady-state flow, interaction between several structures, or structure and a wall, *etc*. In opposite, the added masses theory [4] allows for correct simulation of vibrational-type motion, when the streamlined surface displacement can be neglected, as well as its velocity, and only (vibro-) accelerations are taken into account. The flow in this case is considered quasi-potential. Such approach allows for estimating peak loads, similar to "water hammer" phenomenon, but also does not permit to simulate transient regimes, which are accompanied with intensive vorticity genegation and vortex shedding.

Thus, the only way to correct solving of the above mentioned problem is to simulate the unsteady flow itself, taking into account, if necessary, the interference between several bodies, walls, the pre-history of the structure's motion, *etc*. The main problem is extremal numerical cost of unsteady flow simulation, while the flow itself is, in fact, out of interest – its simulation is required only for hydrodynamical loads estimation. In order to reduce the complexity of simulation, the following simplification is usually assumed: having in mind that in many cases the structures that interact with a flow have large elongation and close to constant cross-section, is can be acceptable to replace spatial flow simulation with considering number of cross-sections of the structure, flow around which can be considered two-dimensional. Thus, the flat fllows around cross-section airfoils are independent "hydrodynamically", but they are coupled implicitly through mutual motions of the cross-sections, if the structure is not rigid. Such simpplification is rather popular in engineering applications and is usually called "flat cross-sections method" and sometimes is mentioned as "quasi-3D" simulation, or "2.5D-simulation".

To simulate the flow, it does not matter – either in "full" 3D statement, or in the framework of flat cross-sections approach, and also for steady hydrodynamical coefficients estimation (and, in fact, for added masses calculation), – the meshless Lagrangian vortex particle methods (VPM) can be used. The review of vortex methods can be found in [5]. VPMs have rather narrow range of applicability that is bounded by incompressible one-phase isothermal flows simulation. However, their numerical complexity can by mush less in comparison to widely-used "classical" mesh-based methods, mainly finite-volume method modifications, implemented in large number of codes. The advantages of vortex methods are manifested most clearly at simulating unbounded external flows (with exact satisfaction of the perturbations decay boundary condition at infinity) and/or Fluid-Structure Interaction problems solving (FSI) with arbitrary translations, rotations, and deformations of the streamlined surface.

However, inspilte of being rather popular in engineering applications, there are only several known freely-available implementations of the vortex particle method, even for two-dimensional flows simulation, that would have been based on modern mathematical models and numerical algorithms, preferably with source code in order to have the possibility to modify it somehow for adaptation to actual problems that required to be solved. In recent years. The only codes `vvflow` [6] and `VM2D` [7] are available, that can be useful for practical purposes; both are based on viscous vortex domains (VVD) method [8]. Note, that the `vvflow` code is written by the developers of the VVD method; the `VM2D` is created by the authors of the present paper; some original algorithms are implemented, mainly connected with improved algorithms of vorticity generation simulation on the airfoil surface.

The aim of this paper is to develop end estimate the efficiency of fast algorithm for vortex particles (that simulate vorticity distribution in the flow domain) mutual interaction simulation and the boundary integral equation solution (that expresses the mathematical model of vorticity generation process).

2. The VM2D code and results of simulation in model problems

The structure and options of the VM2D code are briefly described in [9]. However, the development of the VM2D continues, some other features are available in recent versions, as well as improved algorithms for essential operations. Here let us outline only, that the VM2D code is cross-platform and has a modular structure. In order to improve the performance, parallel algorithms in VM2D are implemented using OpenMP, MPI and Nvidia CUDA technologies, that allows use multiprocessor systems with classical (CPU) architecture and GPU accelerators for computations. In VM2D it is possible to simulate unsteady incompressible flows around airfoil or system of airfoils, calculate hydrodynamic loads acting the airfoils, solve fluid-structure interaction (FSI) problems when the airfoils move under the hydrodynamic loads. It is also possible to simulate internal flows in VM2D.

The results of some two-dimensional model problems solving are presented below, that allows to verify the accuracy and of the implemented algorithms.

2.1. Blasius flow simulation

The Blasius flow [10] seems to be the most popular case for verification of the algorithms for viscous incompressible flows simulation. Flow around semi-infinite thin plate is considered, and exact (asymptotic) solution for it is known. The vortex wake is shown in the figure 1, *a*, and the comparison of the flow velocity profiles in the boundary layer (horizontal and vertical components) in the cross-section $x^* = 0.25L$, where L is plate length in simulation, with exact solution is shown in figure 1, *b*, *c*.

2.2. Poiseuille flow in a flat channel simulation

The Poiseuille flow [10] is considered as an example that allows to verify the VM2D code's capabilities in inner flows simulation. Since there are no tools for the velocity of the flow or pressure setting on some part of the boundary ("inlet" and "outlet" patches), the following approach was implemented: in the beginning and in the ending of the channel number of point sources and sinks, respectively, have been introduced, in order to simulate the required quantity of fluid, being "pumped" through the channel (rate of flow). In figure 2 vortex particles in steady-state regime are shown, as well as velocity profiles in comparison to exact (parabolic) solution at three cross-sections. It is seen, that in middle part of the channel, rather far from sources and sinks, the result is close to analytical one.

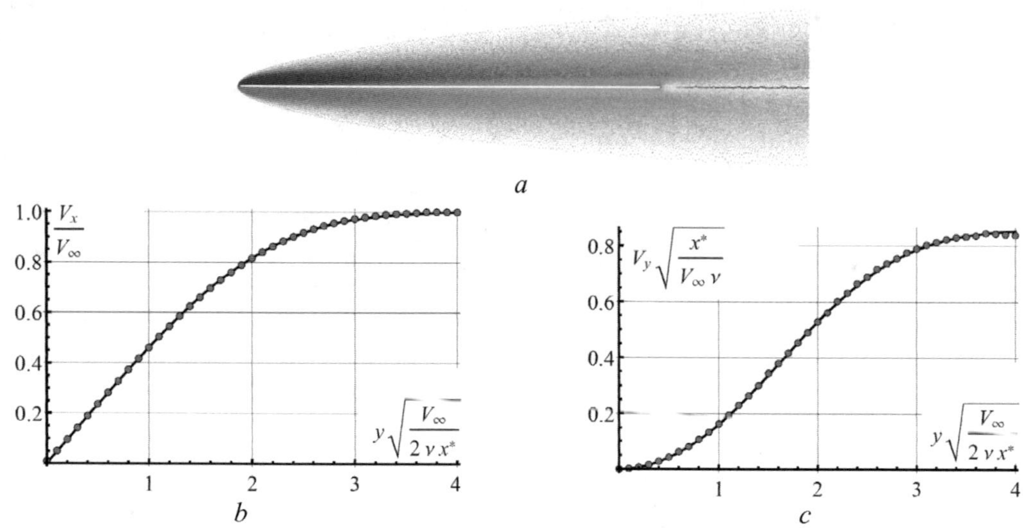

Figure 1. Boundary layer near thin plate (*a*); horizontal (*b*) and vertical (*c*) components of velocity in $x^* = 0.25L$ cross section (red points) in comparison to asymptotic solution (black solid lines).

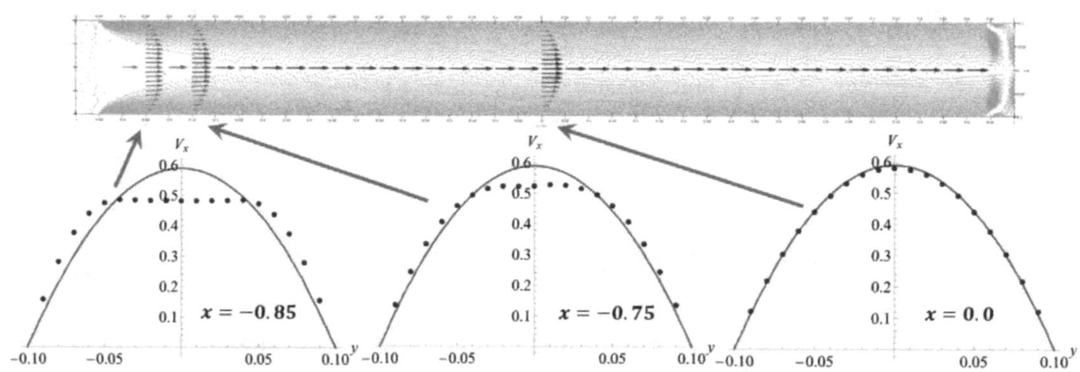

Figure 2. Poiseuille flow in the flat channel simulation. Blue points means horizontal velocities of the flow; red parabolic curve corresponds to the exact solution for given flow rate.

2.3. Couette flow between rotating cylinders simulation

In this case the flow is driven by the walls, that are movable. For the case of flow between circular cylinders rotating with constant angular velocities, exact solution for the steady-state regime is known. The dependencies of rotational velocity component and pressure against radial coordinate are shown in figure 3, *a* and *b*; the distribution of vortices with velocity profile is shown in figure 3, *c*.

2.4. Impulsive start of circular cylinder simulation

The problem of flow simulation and unsteady drag force computation, that acts on impulsively starting cylinder, is one more "classical" test problem, that has evident practical sense. The results of simulation for low values of the Reynolds number Re = 40 and Re = 200 are shown in figure 4. Results for higher value of the Reynolds number Re = 3000 are shown in figure 5.

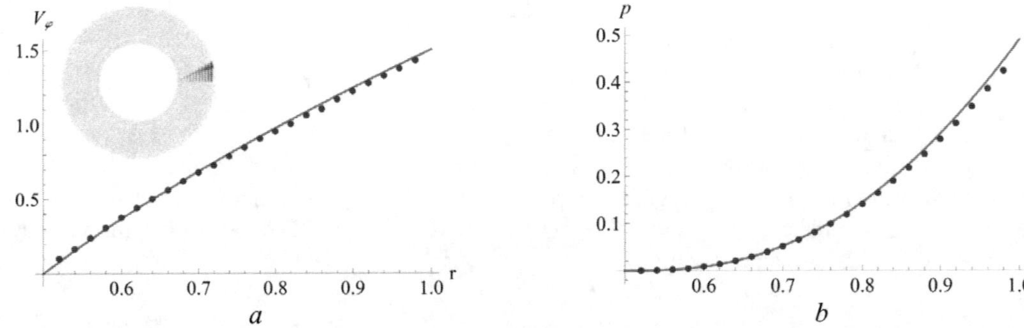

Figure 3. Velocity (*a*) and pressure (*b*) distribution in the gap between rotating cylinders. Blue points correspond to results of numerical simulation in VM2D, red curve is exact analytical solution. Vortex particles fill the gap uniformly (*a*); the inner cylinder assumed to be immovable

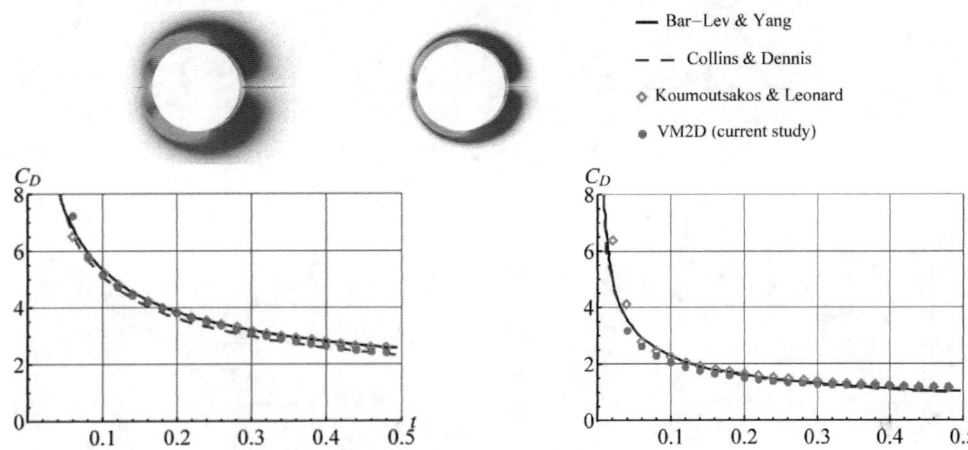

Reynolds number Re = 40 Reynolds number Re = 200

Figure 4. Vortex wake at time moment $t = 0.5$ after the impulsively started cylinders at low Reynolds numbers; unsteady drag coefficient in comparison to results, obtained by other authors [11-13].

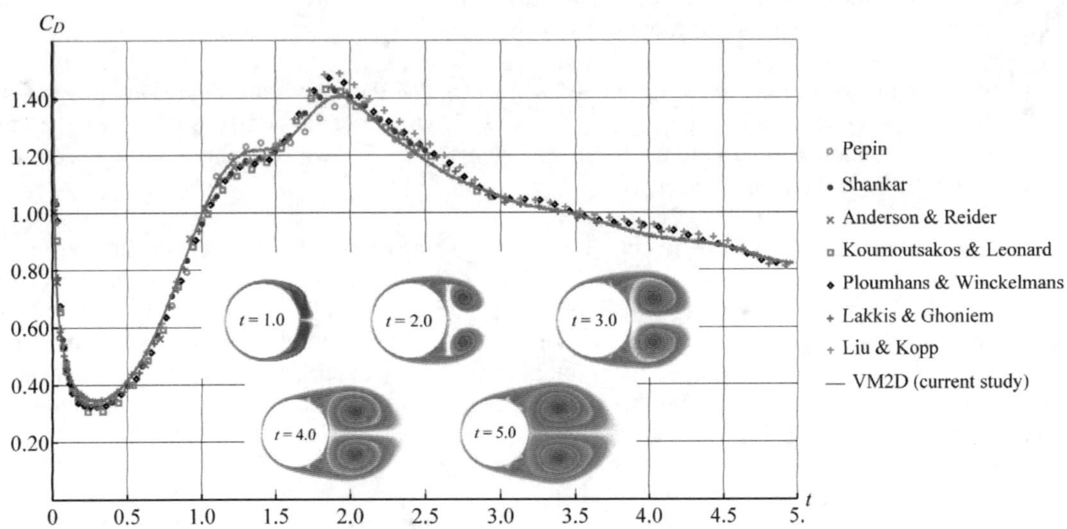

Figure 5. Vortex wake at time moments from $t = 1$ to $t = 5$ after the impulsively started cylinders at Re = 3000 and unsteady drag coefficient in comparison to results, obtained by other authors [13-19].

2.5. Flow around wing airfoil simulation

The VVD method for smooth airfoils allows for correct flow simulation only for low Reynolds numbers, normally not exceeding several thousands. At higher values of Re it is required to take into account somehow the turbulent effects, *i.e.*, three-dimensional character of the flow at small scales, that influences the position of separation point and, in turn, the macroscopic flow parameters.
However, for wing airfoils with sharp edge the position of separation point (as low angles of incidence) is defined by the edge itself, so the VVD method, implemented in the VM2D code, allows for simulation at much higher Reynolds numbers (figure 6).

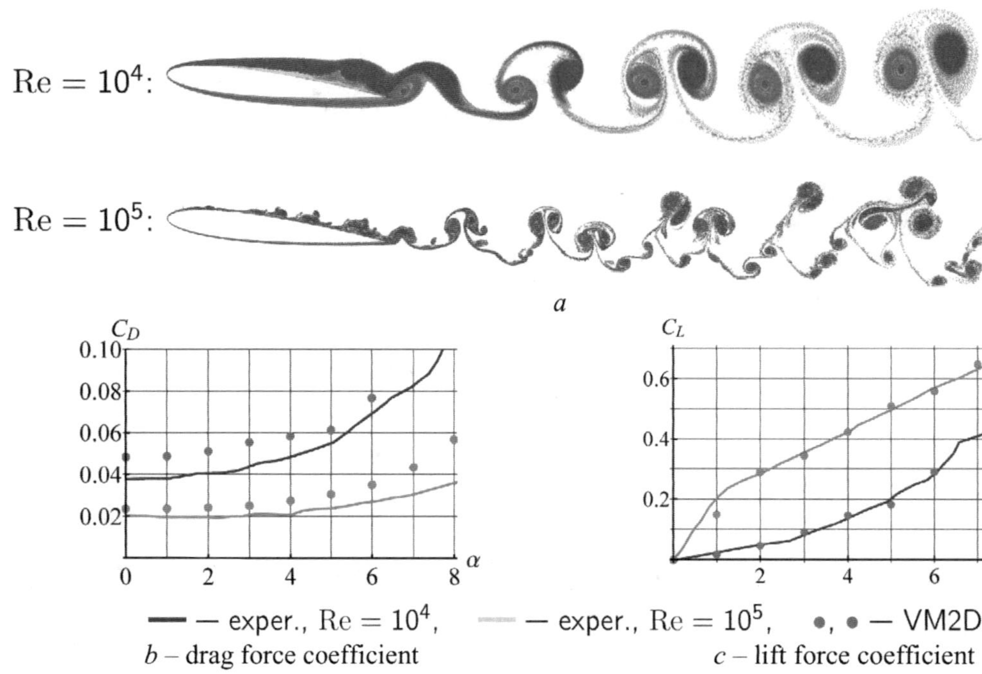

Figure 6. Vortex wake at close to steady-state regime after NACA-0012 wing airfoil at Reynolds numbers Re = 10^4 and Re = 10^5 (*a*); steady drag coefficient (*b*) and lift coefficient (*c*) against angle of incidence in comparison to experimental results.

Number of other model problems, being solved with the VM2D codes, are described in details in [20-22], including not only steady-state regimes, but also dynamical characteristics of vortex shedding, vortex resonance simulation (that belongs to FSI problems), *etc*. So, we can state, that the algorithms implemented in the VM2D code are verified.

3. The numerical complexity of the direct algorithm of vortex particle interaction simulation

The main issue in the first version of the VM2D code is connected to high numerical complexity of the "direct" algorithms that provides evolution simulation of the vortex wake. The convective velocities of vortex particles (that are equal to flow velocity at the corresponding points) are calculated straightforwardly, taking into account all their mutual influences. Namely, according to the Biot – Savart law, the computation of all the velocities (we assume for simplicity that there are no airfoils in the flow domain) for all the vortex particles

$$\mathbf{V}(r_i) = \sum_{j=1}^{N} \Gamma_j \frac{\mathbf{k} \times (\mathbf{r}_i - \mathbf{r}_j)}{2\pi |\mathbf{r}_i - \mathbf{r}_j|^2} \tag{1}$$

requires $O(N^2)$ operations at each time step. At the same time, number of vortex particles in real two-dimensional simulations usually can reach hundreds of thousands and even millions; typical number of time steps has order from thousands to hundreds of thousands, so the direct approach becomes too time-expensive. For example, if one consider the model problem where it is necessary just to compute once velocities of $N = 10^6$ and $N = 2\cdot 10^6$ vortex particles, randomly distributed in some area, it is rather easy to implement direct algorithm, based on formula (1). Time of computations for three, rather modern CPUs and three GPUs is shown in Table 1. Number of cores and streaming multiprocessors is shown; CPU-code was implemented with OpenMP technology, computations are performed with double precision; GPU-code was based on Nvidia CUDA technology, results are given for both single and double precision. Parallelization in both CPU and GPU cases is implemented trivially, over the particles.

Table 1. Time of mutual influence computations by using direct algorithm for $N = 10^6$ and $N = 2\cdot 10^6$ vortex particles on some CPUs and GPUs

	CPUs			GPUs			
	Intel i9-9900	Intel i9-10980XE	2 x Intel Xeon 6254	GeForce 2080 Ti	Titan V	Tesla V100	prec.
Cores / SMs	8	18	32	68	80	80	
10^6 particles	296 s	153 s	151 s	64.8 s	4.8 s	4.7 s	double
				2.0 s	2.1 s	3.2 s	single
$2\cdot 10^6$ particles	1220 s	680 s	640 s	262 s	19.1 s	18.6 s	double
				7.8 s	11.2 s	12.8 s	single

It is clear, that such performance is inacceptable in practice since, recall, typical number of time steps has order $10^4 \dots 10^6$. Sure, additional performance can be achieved by using MPI parallelization technology on clusters with nodes equipped with GPUs, or on nodes with multiple GPUs. For several cluster nodes and several GPUs the speedup is close to linear, but however such a way is connected to involving of enormous computational resources, that is hardly adequate to the problem being solved.
The only way to overcome this issue is turn to use approximate fast methods instead of direct one, based on the Biot – Savart law (1).

4. General idea of tree-based fast methods of the *N*-body problem solving

Since the Biot – Savart law is very similar to gravitational problems, all the approaches, initially developed for *N*-body problems, can be adapted to vortex methods.
The simplest method (from implementational point of view) is the Barnes – Hut (BH) algorithm [23], which provides $O(N \log N)$ numerical complexity. However, significant difference is connected with the fact, that vortex particles can have both positive and negative circulations, while masses of bodies are always positive. The necessary adaptation was performed in [24], but the modified algorithm is not such efficient as initial one. The general idea of the Barnes – Hut method is to split the flow domain (to be more precisely, the domain, containing the vortex particles) hierarchically – *i.e,* to build the *k*-d tree, which cells are considered as clusters of vortex particles. Then, for each vortex particle the tree traversal is performed – far-placed cells (which size is small enough in comparison to distance to them) are considered as one "summary" particle, while that tree leafs, for which the relative distance is small, are processed directly, calculating point-to-point interactions.
The other method, namely Fast Multipole Method (FMM) [25, 26], is based on nearly the same ideas, however quadtree is considered instead of *k*-d tree, and the influence is calculated not for individual particles, but for all the tree cells by constructing the so-called local expansions (which are power series with respect to distance from the tree cell). The influence of each cell is represented as multipole expansion (power series with respect to distance from the influencing cell in negative degrees). Tree traversal is performed only once, particle interactions between neighbouring cells are taken into account straightforwardly. Note, that numerical complexity of the FMM is only $O(N)$.
Numerical experiments show, that classical BH has higher computational cost in comparison to FMM, while FMM does not allows for "smooth" accuracy control: it works perfectly when the required accuracy is high (comparable with machine accuracy), but it is hardly possible to save time of computation by the accuracy reducing (in fact, machine precision is not required in vortex methods).

Such properties of two "basic" approaches allows for developing some hybrid modification – BH/Multipole method, which can be considered as improved version of the BH method: general algorithm remains very similar to classical version of BH, however, some specific features take place:

- tree traversal is performed down to some specified level (let us call the corresponding tree cells as "lowest-level cells"), but not down to the separate particles, which correspond to the leaf-cells in full tree;
- for each cell not only total mass (or circulation) is stored, but also higher multipole moments; which are calculated directly for the leaf-cells (for a cell containing a separate particle only monopole moment is non-zero and equal to its circulation, while all higher-order moments are equal to zero), and by summation of children cells moments, being preliminarily shifted to the centre of the parent cell;
- the influence is calculated for the low-level cells, for which local (Taylor) expansion of the influencing function is constructed, taking into account only those terms in expansion of multipole terms, that has proper order of magnitude; that allows to reduce time of computations by avoiding computing higher-order terms, that do not influence the accuracy;

Note, that in two-dimensional problems such approach is rather efficient, since all the manipulations with multipole terms as well as their local expansions can be written down through complex numbers. As the result, the algorithm is constructed, that contains 3 parameters, which influence the accuracy and numerical complexity of the algorithm (Table 2). Signs "+" and "-" mean that the corresponding influence is essential and can be neglected, respectively.

Table 2. Parameters of the hybrid BH/Multipole method and their influence

Parameter	Accuracy	Complexity
Lowest-level in the tree	–	+
Number of multipoles	+	–
Proximity criterion	+	+

Numerical experiments show, that only the optimal depth of the tree (more precisely, the depth of "traversable" part of the tree, *i.e.*, the above mentioned depth of lowest-level cells) depends significantly on number of particles, it is proportional to $\log_2 N$; if vortex particles fill rectangular region with high aspect ratio, the optimal depth should be additionally increased by $\log_2 \frac{a}{b}$, where a and b are maximal and minimal side lengths. Number of multipole terms and proximity criterion are chosen to satisfy the accuracy requirements and provide minimal time of computations.

Let us note also that *k*-d tree construction is logically rather simple operation, however the efficiency of its implementation influences significantly the efficiency of the whole algorithm.

For example, if we implement the "naive" algorithm of the tree construction, its numerical complexity, caused by large number of operations (mainly – comparisons) leads to the following result: tree construction subroutine takes around 10 % of the whole algorithm, and the scalability of this "naive" algorithm is rather poor: when using CPU with 32 cores its speedup is only about 5 while speedup of other procedures is over 25, so tree construction now takes 35 % of the algorithm.

Such issue is overcome by implementing the Morton-tree algorithm, based on fractal z-curve. Its construction does not require operations with pointers, since the tree is stored in linearly indexed array. In the Table 3 the time of computations is shown, that is required for velocities computation for $N = 2 \cdot 10^6$ vortex particles by using the developed BH/multipole algorithm with optimal choise of all above mentioned parameters. Three different accuracy levels ε are considered. Computations are performed on the following CPUs: 2xIntel Xeon 6254 Gold (32 cores) and Intel i9-10980XE (18 cores).

Table 3. Time of computations (in seconds) on CPUs

Accuracy	2xIntel Xeon 6254 Gold (32 cores)			Intel i9-10980XE (18 cores)		
	Tree	Upward	Downward	Tree	Upward	Downward
10^{-3}	0.075	0.031	0.119	0.094	0.055	0.198
10^{-5}	0.077	0.044	0.221	0.097	0.079	0.386
10^{-7}	0.075	0.071	0.315	0.096	0.131	0.511

Here "Upward" means computation of all the tree parameters (multipole moments for all the cells), "Downward" means tree traversals that allows for obtaining local expansions in the lowest-level cells and their usage for vortex particles' velocities computation.
The implemented algorithm is good enough scalable, the speedup values for all operations are given in the Table 4 together with estimation of parallelization efficiency.

Table 4. Speedup of the operations on hybrid BH/multipole method

Accuracy	2xIntel Xeon 6254 Gold (32 cores)				Intel i9-10980XE (18 cores)			
	Tree	Upward	Downward	Efficiency	Tree	Upward	Downward	Efficiency
10^{-3}	11.2	26.5	28.1	*0.69*	7.8	13.0	14.7	*0.70*
10^{-5}	11.1	27.4	29.4	*0.78*	7.6	13.2	14.8	*0.74*
10^{-7}	11.7	28.9	29.7	*0.83*	7.8	13.2	15.8	*0.79*

The developed algorithm was adapted for GPUs, however, some modifications have been implemented:

- the full quad-tree is constructed instead of *k*-d tree;
- control points are not grouped into clusters (*i.e.*, the tree traversal is performed for all the tree leafs), thus, local expansions are not constructed.

The known implementation of the Barnes – Hut method was used [27], however it was modified significantly.
Time of computations for above mentioned rather powerful GPUs in comparison with CPUs for the same problem with $N = 2\cdot 10^6$ vortex particles with three levels of accuracy is shown in the Table 5. Time of computations on GPUs is given for single and double precision; for comparison the last line contains the results for direct computations (in seconds, without fast method). The efficiency of the fast algorithm is evident.

Table 5. Time of computations on CPUs and GPUs (in milliseconds)

Accuracy	CPUs			GPUs		
	Intel i9-9900	Intel i9-10980XE	2 x Intel Xeon 6254	GeForce 2080 Ti	Titan V	Tesla V100
	8 cores	18 cores	32 cores	68 SM	80 SM	80 SM
10^{-3}	433	348	226	180 / 23	29 / 23	26 / 21
10^{-5}	605	563	345	332 / 37	49 / 35	51 / 33
10^{-7}	773	739	463	586 / –	75 / –	90 / –
direct	1220 s.	680 s.	640 s.	262 / 7.8 s.	19.1 / 11.3 s.	18.6 / 12.8 s.

5. Linear system solving by using fast method

Another time-consuming operation in vortex methods is the boundary integral equation (BIE) solution with respect to vorticity intensity that is generated on the airfoil surface in the flow. By using some numerical scheme [28] the linear system – discrete analogue of the BIE is constructed. It can be solved

directly by using Gaussian elimination method, or iteratively, by using, for example, GMRES method (with preconditioning). It can be easily shown, that in iterative method matrix-to-vector multiplication [29] is similar to computation of velocity of vortex particle, with only difference, that the velocity is induced not by point vortices, as earlier, but by the panels (parts of the surface line) with vorticity distributed over them. In the most accurate scheme with rectilinear panels vorticity distribution is piecewise-linear. Note, that application of the fast method allows not to calculate and store all the matrix components; normally not more than 0.1 % of matrix components are calculated explicitly.
The results of numerical experiments for the airfoil, which surface line is split into N = 3 200 and N = 12 800 panels for sequential and parallel algorithms are shown in Table 6. The error provided by the approximate fast method is chosen to be of the same order as one for the numerical scheme of the BIE approximation. Direct linear system solution have been performed by using Intel MKL library; GMRES was implemented with straightforward matrix-to-vector multiplication and with developed hybrid BH/Multipole algorithm with optimal choice of parameters.

Table 6. Time of the BIE solution (in seconds), CPU Intel i9-10980XE, 18 cores

Method	N = 3 200		N = 12 800	
	Sequential	Parallel	Sequential	Parallel
Gauss, Intel MKL	6.59	0.715	311	16.5
GMRES, multipl.	4.82	0.391	82.2	6.65
GMRES, fast method	0.10	0.015	0.711	0.072

The preconditioner is quite important since it allows for twice reduction of number iterations, the preconditioner matrix has tri-diagonal structure; modification of the sweep method is developed having linear numerical complexity.

6. Conclusion

A modification of the fast Barnes – Hut algorithm is developed using the ideas of the multipole method for calculating the velocities of vortex elements; the efficient parallel implementation of fast algorithm are developed using OpenMP and Nvidia CUDA technologies: dozens of times speedup is achieved in comparison to sequential code. The method is adapted for solving the boundary integral equation; higher-order numerical schemes are supported. The following features of the algorithm should be pointed out: quasilinear $O(N \log N)$ computational complexity; no need to store the full matrix of the system; efficient preconditioning technique; possibility of generalization for three-dimensional case.
The developed codes are available on GitHub: https://github.com/vortexmethods/fastm

References

[1] Cottet G-H and Koumoutsakos P 2000 *Vortex methods: theory and practice* (Cambridge: CUP)
[2] Lewis R I 1991 *Vortex element methods for fluid dynamic analysis of engineering systems* (Cambridge: CUP)
[3] Branlard E 2017 *Wind turbine aerodynamics and vorticity-based methods: Fundamentals and recent applications* (Berlin: Springer)
[4] Korotkin A I 2009 *Added Masses of Ship Structures* (Berlin: Springer)
[5] Mimeau C and Mortazavi I 2021 A review of vortex methods and their applications: from creation to recent advances *Fluids* **6** 68 doi 10.3390/fluids6020068
[6] Vvflow CFD suite https://github.com/vvflow/vvflow
[7] VM2D: Vortex method for 2D flow simulation https://github.com/vortexmethods/VM2D
[8] Dynnikova G Ya 2004 The Lagrangian approach to solving the time-dependent Navier-Stokes equations *Doklady Physics* **49:11** pp 648-652 doi 10.1134/1.183153

[9] Marchevsky I, Sokol K, Ryatina E and Izmailova Y 2023 The VM2D open source code for two-dimensional incompressible flow simulation by using fully Lagrangian vortex particle methods *Axioms* **12** 248 doi 10.3390/axioms12030248
[10] Rosenhead L 1963 *Laminar boundary layers* (Oxford: Clarendon Press)
[11] Bar-Lev M and Yang H T 1975 Initial flow over an impulsively started circular cylinder *J Fluid Mech.* **72** pp 625-647 doi 10.1017/S0022112075003199
[12] Collins W M and Dennis S C R 1973 The initial flow past an impulsively started circular cylinder *Q. J. Mech. App. Math.* **26** pp 53-75 doi 10.1093/qjmam/26.1.53
[13] Koumoutsakos P and Leonard A 1995 High-resolution simulations of the flow around an impulsively started cylinder using vortex methods *J Fluid Mech* **296**, pp 1-38 doi 10.1017/S0022112095002059
[14] Pepin F M 1990 Simulation of the flow past an impulsively started cylinder using a discrete vortex method *PhD Thesis*, California Institute of Technology, Pasadena, CA, USA
[15] Shankar S A 1996 New mesh-free vortex method *PhD Thesis*, FAMU-FSU College of Engineering, Tallahassee, FL, USA
[16] Anderson C B and Reider M B 1996 A high order explicit method for the computation of flow about a circular cylinder *J. Comput. Phys.* **125** pp 207-224 doi 10.1006/jcph.1996.0089
[17] Ploumhans P and Winckelmans G S 2000 Vortex methods for high-resolution simulations of viscous flow past bluff bodies of general geometry *J. Comput. Phys.* **165** pp 354-406 doi 10.1006/jcph.2000.6614
[18] Lakkis I and Ghoniem A 2009 A high resolution spatially adaptive vortex method for separating flows. Part I: Two-dimensional domains *J. Comput. Phys.* **228** pp 491-515 doi 10.1016/j.jcp.2008.09.025
[19] Liu Z and Kopp G A 2011 High-resolution vortex particle simulations of flows around rectangular cylinders *Comp. Fluids* **40** pp 2-11 doi 10.1016/j.compfluid.2010.07.011
[20] Kuzmina K, Marchevsky I, Soldatova I and Izmailova Y 2021 On the scope of Lagrangian vortex methods for two-dimensional flow simulations and the POD technique application for data storing and analysing *Entropy* **23** 118 doi 1 0.3390/e23010118
[21] Kuzmina K and Marchevsky I 2021 Flow simulation around circular cylinder at low Reynolds numbers using vortex particle method *J. Phys.: Conf. Ser.* **1715** 012067 doi 10.1088/1742-6596/1715/1/012067
[22] Marchevsky I K, Kuzmina K S and Ryatina E P 2020 Flow Simulation around a circular cylinder for low Reynolds numbers and determining the flow separation point in VM2D code *Proc. Topical Problems of Fluid Mechanics* pp 144-151 doi 10.14311/TPFM.2020.019
[23] Barnes J and Hut P 1986 A hierarchical $O(N \log N)$ force-calculation algorithm *Nature* **324:4** pp 446-449 doi 10.1038/324446a0
[24] Dynnikova G Y 2009 Fast technique for solving the N-body problem in flow simulation by vortex methods *Computational Mathematics and Mathematical Physics* **49:8** pp 1389-1396 doi 10.1134/S0965542509080090
[25] Greengard L 1988 The rapid evaluation of potential fields in particle systems *PhD thesis* Yale University p 101
[26] Greengard L and Rokhlin V 1987 A fast algorithm for particle simulations *J. Comput. Phys.* **73:2** pp 325-348 doi 10.1016/0021-9991(87)90140-9
[27] Burtscher M and Pingali K 2011 An Efficient CUDA Implementation of the Tree-based Barnes Hut n-Body Algorithm *GPU Computing Gems Emerald Edition* pp 72-91 doi 10.1016/B978-0-12-384988-5.00006-1
[28] Marchevsky I K, Sokol K S and Izmailova Yu A 2022 T-schemes for mathematical modelling of vorticity generation on smooths airfoils in vortex particle methods *Herald of the Bauman Moscow State Tech. Univ. Ser. Natural Sci.* **6** pp 33-59 doi 10.18698/1812-3368-2022-6-33-59
[29] Mikhailov E A, Marchevskii I K and Kuzmina K S 2019 Iterative approach to solving boundary integral equations in the two-dimensional vortex methods of computational hydrodynamics *J. Appl. Industr. Math.* **13:4** pp 672-684 doi 10.1134/S1990478919040100

On the locating rainbow connection number of amalgamation of complete graphs

A W Bustan[1,3,*], A N M Salman[2], P E Putri[2]

[1] Doctoral Program of Mathematics, Faculty of Mathematics and Natural Sciences, Institut Teknologi Bandung, Bandung, Indonesia
[2] Combinatorial Mathematics Research Group, Faculty of Mathematics and Natural Sciences, Institut Teknologi Bandung, Bandung, Indonesia
[3] Mathematics Department, Faculty of Mathematics and Natural Sciences, Universitas Pasifik Morotai, Indonesia

* Corresponding author's e-mai: ariestha.bustan@students.itb.ac.id

Abstract. Locating rainbow connection number determines the minimum number of colors connecting any two vertices of a graph with a rainbow vertex path and also verifies that the given colors produce a different rainbow code for each vertex. Locating rainbow connection number of graphs is a new mathematical concept, especially in graph theory, which combines the concepts of the rainbow vertex coloring and the partition dimension. In this paper, we determine the locating rainbow connection number of amalgamation of complete graphs.

1. Introduction

Let G be a connected, undirected and simple graph with order n. For $k \in N$, the color of vertices of G is called a *rainbow vertex k-coloring* if there is a function $c: V(G) \to 1,2,\ldots,k$ such that every two distinct vertices $x, y \in V(G)$ there is *rainbow vertex* $x - y$*-path* whose all other vertices on the path besides vertex x and vertex y have a different color. The *rainbow vertex connection number* of G, which is represented by $rvc(G) = min\{k: k \in N\}$, such that G has a rainbow vertex k-coloring [1]. For example, the rainbow vertex connection numbers of simple graph class formed by graph operation are given in [2] and [3].

Let R_i be the set of vertices with color i, for $i \in \{1, 2, \ldots, k\}$ and $\Pi = \{R_1, R_2, \ldots, R_k\}$be an ordered partition of $V(G)$. Write $rc_\Pi(v) = \left(d(v, R_1), d(v, R_2), \ldots, d(v, R_k)\right)$. Where $d(v, R_i) = min\{d(v, y): y \in R_i\}$ for each $i \in \{1,2,\ldots,k\}$. Further, we call $rc_\Pi(v)$ the *rainbow code* of v of G in relation to Π. If $rc_\Pi(v_j) \neq rc_\Pi(v_l)$ for distinct $j, l \in \{1,2,\ldots,n\}$ then c is known as the *locating rainbow k-coloring* of G. $rvcl(G) = min\{k: k \in N\}$ expression for G's locating rainbow connection number indicates that G has a locating rainbow k-coloring. Every locating rainbow k-coloring of G is a rainbow vertex coloring of G, giving us $rvc(G) \leq rvcl(G)$. Based on an easy observation by Yuster and Krivelevich, $rvc(G)$ is more than or equals to the diameter minus one, so that we have $rvcl(G) \geq diam - 1$.

Several results related to the upper and lower bounds of locating rainbow connection numbers have been obtained in [4]. In Lemma 1.1, two vertices with the same distance from other vertices in G must be assigned a different color so that the rainbow codes for the two vertices are distinct. Besides that, in Lemma 1.2, for a graph G of order at least three, we have $rvcl(G)$ is between 2 and the order of G. These results are presented in the following lemma.

Lemma 1.1 [4] *Let c be a locating rainbow coloring of G. Let j and k be two distinct vertices of G. If* $d(j,x) = d(k,x)$ *for all* $x \in V(G) - j, k$, *then* $c(j) \neq c(k)$.

Lemma 1.2 [4] *Let n be a positive integer* with $n \geq 3$. *If G is a connected graph of order n, then*

$$2 \leq rvcl(G) \leq n.$$

Together with the two lemmas mentioned previously, we have also calculated the lower bound by counting the pendant vertices. The maximum number of pendants adjacent to one vertex in G is the smallest number of colors that can be used for a graph with pendants so that each vertex has a unique rainbow code. We have also obtained locating rainbow connection numbers from several simple graph classes, such as paths, stars, and complete graphs. For paths with order at least three, most of the locating rainbow connection number of paths equals the diameter minus one, except for path graphs with order two and order three, which have the locating rainbow connection number equal to diameter. Figure 1 shows the rainbow codes of P_2, P_3, and P_4.

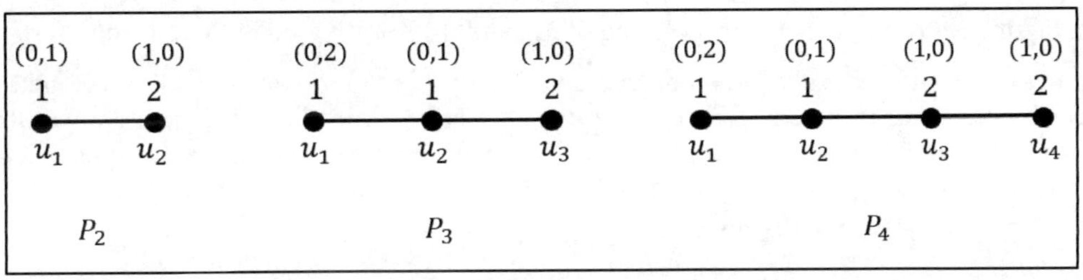

Figure 1. Locating rainbow coloring of P_2, P_3, and P_4

Regarding the star graph S_m of order $m + 1$, $rvcl(S_m)$ is differs significantly from $rvc(S_m)$; in the star graph, $rvc(S_m)$ equals the number of vertices (one), while $rvcl(S_m)$ equals the number of pendant vertices (the order of a star minus one). Figure 2 shows rainbow vertex coloring of S_8 and the rainbow codes of S_8

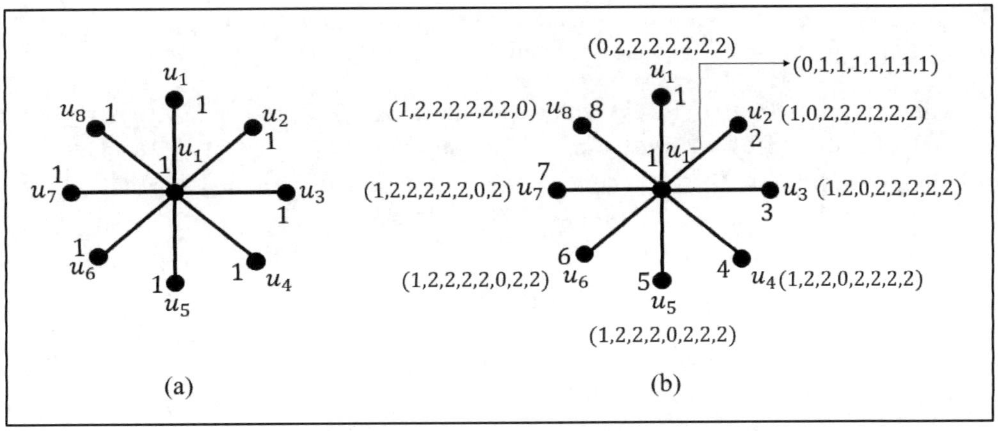

Figure 2. (a) Rainbow vertex coloring of S_8, and (b) Locating rainbow coloring of S_8

Same with the star graph, the complete graph K_n of order n also has a rainbow vertex connection number, which is very different from the locating rainbow connection number. The complete graph has a rainbow vertex connection number equal to 1, while the locating rainbow connection number of complete graphs equals the order of the graph. The following theorem shows this result.

Theorem 1.1 *Let n be a positive integer with* $n \geq 3$ *and* K_n *be a complete graph of order* n. *Then the locating rainbow connection number of* K_n *is* $rvcl(K_n) = n$.

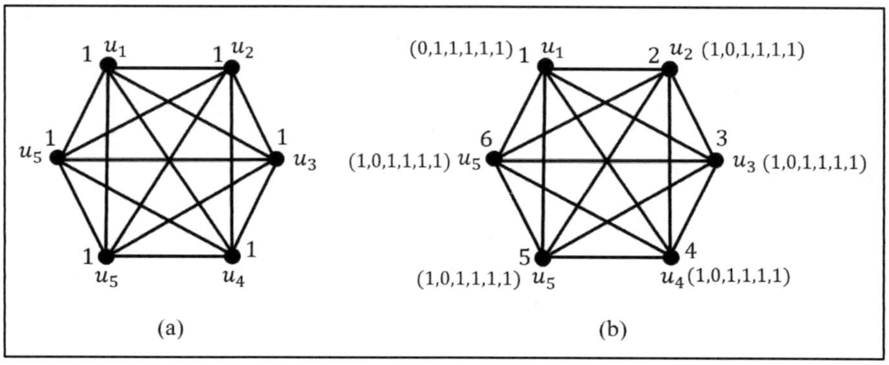

Figure 3. (a) Rainbow vertex coloring of K_6, and (b) Locating rainbow coloring of K_6

We use Lemma 1.1 to prove that a complete graph cannot be colored with $n-1$ colors. Next, we get the upper bound by coloring all the vertices a different color. After looking for rainbow connection numbers in several classes of graphs, we started to explore operations on graphs, starting with the amalgamation of a graph. Several researchers have previously studied the concept of rainbow coloring on amalgamation of graphs in [5] and [6].

In 2016, Rosmaini and Salman provided the rainbow vertex connection number of the amalgamation of some graphs, but this was not applied to the complete graph because the diameter of the amalgamation of the complete graph is two, and based on easy observation the rainbow vertex connection number is trivial. On the other hand, these trivial results are not obtained for locating rainbow connection number for complete graphs. Therefore, we demonstrate the amalgamation of complete graphs' locating rainbow connection number in the following chapter

2. Main results

For an integer $t \geq 1$, let $\{G_1, G_2, \dots, G_t\}$ be a collection of connected, finite, and simple graph and G_i contains a fixed vertex v_0, also known as the central vertex. *Amalgamation* of $G_1, G_2, \dots, G_t$ denoted by $Amal(G, v_0, t)$ is a graph obtained by attaching all the central vertices of every vertex in the graph G_i.

Next, we will show $rvcl(Amal(G, v_0, t))$ with $G \cong K_n$, where K_n be a complete, n-order graph. Define the set of edges and vertices of $Amal(K_n, v_0, t)$ are as follows. $E(Amal(K_n, v_0, t)) = \{v_0 v_{i,j} | i \in [1,t], j \in [1, n-1]\} \cup \{v_{i,j} v_{i,k} | i \in [1,t], j \in [1,n-1], k \in [1,n-1]\}$ and $V(Amal(K_n, v_0, t)) = \{v_0\} \cup \{v_{i,j} | i \in [1,t], j \in [1,n-1]\}$. (See Figure 4).

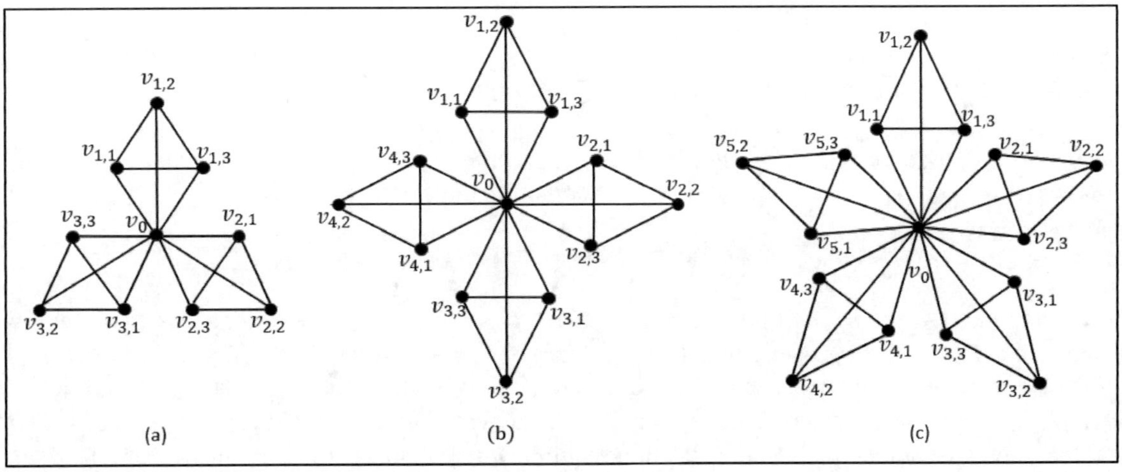

Figure 4. (a) $\boldsymbol{Amal(K_4, v_0, 3)}$, (b) $\boldsymbol{Amal(K_4, v_0, 4)}$, and (c) $\boldsymbol{Amal(K_4, v_0, 5)}$.

Based on the sets of vertex and edge above we have $d(v_{i,j}, x) = d(v_{i,k}, x)$ for distinct, $j, k \in [1, n-1]$ and $i \in [1, t]$ for every $x \in V(Amal(K_n, v_0, t))$, such that by Lemma 1.1, we have Corollary 2.1 below.

Corollary 2.1 *Let n and t be two integers where $n \geq 3$ and $t \geq 1$. Let c be a locating rainbow coloring of $Amal(K_n, v_0, t)$. Then $c(v_{i,j}) \neq c(v_{i,k})$ for distinct $j, k \in [1, n-1]$, $i \in [1, t]$, for every $x \in V(Amal(Kn, v_0, t))$.*

We have diameter of $Amal(K_n, v_0, t)$ is 2. Every collection K_i of $Amal(K_n, v_0, t)$ contains the following vertex and edge sets. $V(K\_i) = \{v_0\} \cup \{v_{i,j} | i \in [1, t], j \in [1, n-1]\}$ and $E(K_i) = \{v_0 v_{i,j} | i \in [1, t], j \in [1, n-1]\} \cup \{v_{i,j} v_{i,k} | i \in [1, t], j \in [1, n-1], k \in [1, n-1]\}$. Therefore, we get $d(v_{i,j}, v_0) = d(v_{k,l}, v_0) = 1$ for $i, k \in [1, t], j, l \in [1, n-1]$ and $d(v_{i,j}, v_{s,t}) = d(v_{k,l}, v_{s,t}) = 2$, where $i \neq k \neq s$ for $i, k, s \in [1, t]$ and $j, l, t \in [1, n-1]$. Thus, the following corollary is hold.

Corollary 2.2 *Let n and t be two integers where $n \geq 3$ and $t \geq 2$. Let A be the set of colors assigned to the vertices in $V(K_i)$and B be the set of colors assigned to the vertices in $V(K_k)$ for $i \neq k$ and $i, k \in [1, t]$. If c is a locating rainbow coloring of $Amal(K_n, v_0, t)$, then $A \neq B$.*

We show a lemma that provides the lower bound of $rvcl(Amal(K_n, v_0, t))$ by applying $rvcl(K_n)$ before determining $rvcl(Amal(K_n, v_0, t))$.

Lemma 2.1. *Let n and t be two integers where $n \geq 3$ and $t \geq 1$. Let K_n be a complete graph of n order, v_0 be a central vertex of K_n, and t be a number of collections of K_n. Then*

$$rvcl(Amal(K_n, v_0, t)) \geq rvcl(K_n)$$

Proof. For $t = 1$, based on Theorem 1.1, we have $rvcl(Amal(K_n, v_0, t)) = n$. Next, for $t \geq 2$, suppose that $rvcl(Amal(K_n, v_0, t)) \leq rvcl(K_n) - 1 = n - 1$. Without loss of generality, assign color 1 to vertex v_0. Each collection K_i for $i \in [1, t]$ has $n - 1$ vertices with the same neighbour to other vertices in $Amal(K_n, v_0, t)$. Therefore, by Corollary 2.2, $n - 1$ of these vertices must be given $n -$ 1distinct colors. Therefore, without loss of generality, consider collections K_2 and K_1 and assign colors $1, 2, \ldots, n-1$ to the vertex sets $\{v_{2,1}, v_{2,2}, \ldots, v_{2,n-1}\}$ and $\{v_{1,1}, v_{1,2}, \ldots, v_{1,n-1}\}$ respectively. Contradiction with Corollary 2.2. Thus, $rvcl(Amal(K_n, v_0, t)) \geq rvcl(K_n)$.

The following theorem determines $rvcl\big(Amal(K_n, v_0, t)\big)$ by observing the values of t that satisfy the equation if $rvcl\big(Amal(K_n, v_0, t)\big)$ and the order of $Amal(K_n, v_0, t)$ are known.

Theorem 2.1. *Let n and t be two integers where n ≥ 3 and t ≥ 2. Let K_n be a complete graph of n order, v_0 be a central vertex of K_n, and t be a number of collections of K_n . Then*

$$rvcl(Amal(K_n, v_0, t)) = r, \text{ if } C(r-1, n-1) + 1 \le t \le C(r, n-1)$$

Given that $rvcl\big(Amal(K_n, v_0, t)\big) = r$, by Lemma 2.1, we have $r \ge n$. Without loss of generality, assign color 1 to vertex v_0. Next, based on Corollary 2.1, $n-1$ vertices in $V(K_i)/\{v_0\}$ for $i \in [1, t]$ must have a different color and based on Corollary 2.2, the set of colors that assign to vertices in $V(K_i)/\{v_0\}$ cannot be the same as the set of colors that assign to the vertices in $V\big(K_j\big)/v_0$ for $, j \in [1, t]$ and $i \ne j$. As a result, the maximum collection required with r colors is the combination of r from $n-1$ or $C(r, n-1)$, so that $t \le C(r, n-1)$ is obtained and we also get $t \ge C(r-1, n-1) + 1$.
We have the diameter of $Amal(K_n, v_0, t)$ is 2, so for every two vertices in $Amal(K_n, v_0, t)$, there is always a path whose internal vertices are colored differently connecting them by using r colors. Furthermore, by assigning a different color to each vertex in $V(K_i)/\{v_0\}$ and a different set of colors in each K_i as stated in Corollary 2.2, each collection must have at least one color that is not used, which causes each vertex in the collection to have a distance equal to two from that color and each vertex has a unique rainbow code (see Figure 5 for an illustration).

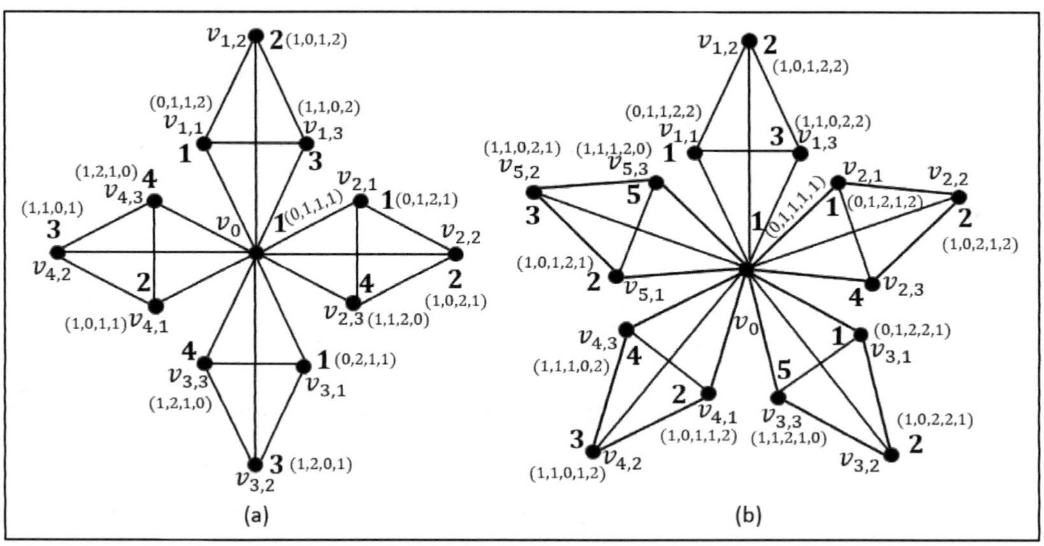

Figure 5. Rainbow codes of (a) $\boldsymbol{Amal(K_4, v_0, 4)}$ and (b) $\boldsymbol{Amal(K_4, v_0, 5)}$.

Acknowledgments
This research was supported by LPDP from Ministry of Finance Indonesia and the Ministry of Education, Culture, Research, and Technology.

References

[1] M. Krivelevich and R. Yuster, "The rainbow connection of a graph is (at most) reciprocal to its minimum degree," *J. Graph Theory*, vol. 63, no. 3, pp. 185–191, 2010.

[2] A. W. Bustan and A. N. M. Salman, "The rainbow vertex connection number of star wheel graphs," *AIP Conf. Proc.*, vol. 2202, no. December, 2019.

[3] D. N. S. Simamora and A. N. M. Salman, "The Rainbow (vertex) connection number of pencil graphs," *Procedia Comput. Sci.*, vol. 74, pp. 138–142, 2015

[4] Bustan, A. W., Salman, A. N. M., & Putri, P. E. (2021, February). On The Locating Rainbow Connection Number of A Graph. In *Journal of Physics: Conference Series* (Vol. 1764, No. 1, p. 012057). IOP Publishing

[5] Awanis, Z. Y., & Salman, A. N. M. (2022). The strong 3-rainbow index of some certain graphs and its amalgamation. *Opuscula Mathematica*, *42*(4), 527-547.

[6] Fitriani, D., & Salman, A. N. M. (2016). Rainbow connection number of amalgamation of some graphs. *AKCE International journal of graphs and combinatorics*, *13*(1), 90-99.

Coefficient estimation of simultaneous equations in the case of stochasticity of exogenous variables

A M Minitaeva

Department of Computer Systems and Networks, Bauman Moscow State Technical University, 5/1 ul. Baumanskaya 2-ya, Moscow 105005, Russian Federation

Corresponding author's e-mail: minitaeva@bmstu.ru

Abstract. The paper deals with the question of obtaining such an estimate of the parameters of the model of simultaneous equations, which would have better statistical properties in comparison with the estimates of both the usual method of least squares and two-, three-step methods of least squares, as well as the method of fixed point and other methods of evaluating the coefficients of the equations considered in the system simultaneously. The main drawback of methods based on minimization of the sum of squares of deviations of real observations from some theoretical dependence is that the sum of squares of errors that correlate with the values of variables included in the expression for these errors is minimized. A violation of the basic assumption of least-squares methods about the independence of predetermined variables and errors leads to the fact that least-squares estimation methods in this case lose their best properties: unbiasedness, efficiency, and consistency. The paper proves the theorem about the efficiency of estimates obtained using two-step least squares methods and that of estimates obtained using the method of factorization of simultaneous equations.

1. Introduction

In contrast to a set of regression equations that have a unidirectional causal relationship from independent variables to the resulting indicators in systems analysis, there are equations with differently directed explanatory effects of simultaneous variables. Such variables are interdependent or endogenous variables. A unique property of systems of simultaneous equations is that an endogenous variable - a regressor in "one's" equation - can appear in another as a regressor of an explanatory variable. And as a consequence, endogenous variables become correlated with random perturbations of the equations [1,2]. Their number in such a system should be equal to the number of endogenous variables.

The usual least squares method cannot be used to obtain estimates of the parameters of the simultaneous equations model. Thus, there is a need for other estimation methods that can circumvent the problem of correlated variables and errors.

2. Materials and methods

Here we write down the general (behavioral) form of the system of simultaneous equations:

$$y_{t1} = \beta_{21}y_{t2} + \beta_{31}y_{t3} + \dots + \beta_{k1}y_{\mathrm{tk}} + \gamma_{11}x_{t1} + \gamma_{21}x_{t2} + \dots + \gamma_{n1}x_{\mathrm{tn}} + u_{t1};$$

$$y_{t2} = \beta_{12}y_{t1} + \beta_{32}y_{t3} + \dots + \beta_{k2}y_{\mathrm{tk}} + \gamma_{12}x_{t1} + \gamma_{22}x_{t2} + \dots + \gamma_{n2}x_{\mathrm{tn}} + u_{t2};$$

$$\dots\dots\dots \quad (1)$$

$$y_{tk} = \beta_{1k}y_{t1} + \beta_{k2}y_{t2} + \dots + \beta_{k-1,k}y_{t,k-1} + y_{1k}x_{t1} + y_{2k}x_{t2} + \dots + y_{n2}x_{tn} + u_{tk};$$

where y_{ti} is the value of the i-th endogenous variable $Y_i(i=1,2,\dots,n)$ at time t; k is the number of endogenous variables, T is the number of observations; x_{tl} is the value of the i-th exogenous variable X_l $(l=1,2,\dots,n)$ at time t $(t=1,2,\dots T))$; β_{ij} is the coefficient at endogenous variable Y_{in} i-th equation of the system; Y_{lj} is the coefficient at exogenous variable in i-th equation of the system; u_{tj} is random deviation of i-th equation of the system at time t.
It should be noted that for a particular system of equations (1) not every variable both endogenous and exogenous should be included in every equation. This is determined by the economic sense of the system itself and the identification criteria of the equations of the system [3]. The system of equations (1) can be represented in matrix form. Let us introduce notations:

- k – is the number of endogenous variables and, consequently, the number of equations in the simultaneous equation model;
- n – is the number of exogenous variables;
- $y_t = (y_{t1} \quad y_{t2} \quad \cdots \quad y_{tk})$ – is the vector of values of endogenous variables;
- $x_t = (x_{t1} \quad x_{t2} \quad \cdots \quad x_{tn})$ – is the vector of values of exogenous variables;
- $u_t = (u_{t1} \quad u_{t2} \quad \cdots \quad u_{tk})$ – is the vector of random deviations of equations;
- $B = \begin{pmatrix} 1 & -\beta_{12} & \dots & -\beta_{1k} \\ -\beta_{21} & 1 & \dots & -\beta_{2k} \\ \dots & \dots & \dots & \dots \\ -\beta_{k1} & -\beta_{k2} & \dots & 1 \end{pmatrix}$ – is the matrix of coefficients with endogenous variables; on its main diagonal there are ones, which reflects one of the features of the system of simultaneous equations: for each endogenous variable there is its own explanatory equation, in which this variable is on the left side of the equation;
- $-r = \begin{pmatrix} y_{11} & y_{12} & \dots & y_{1k} \\ y_{21} & y_{22} & \dots & y_{2k} \\ \dots & \dots & \dots & \dots \\ y_{n1} & y_{n2} & \dots & y_{nk} \end{pmatrix}$ – is the matrix of coefficients at exogenous variables;
- $I = \begin{pmatrix} 1 & 0 & \dots & 0 \\ 0 & 1 & \dots & 0 \\ \dots & \dots & \dots & \dots \\ 0 & 0 & \dots & 1 \end{pmatrix}$ – is the identity matrix (with ones on the main diagonal, zeros on other positions).

A "minus" sign before the coefficients of endogenous variables and before the matrix of coefficients of exogenous variables will be required for the structural form of the system of equations. Capital non-bold letters with one index denote variable names, the index corresponds to the variable number; lowercase non-bold letters with two indices denote the numerical values of the corresponding variables, the first index corresponds to the time moment, the second to the variable number; bold capital letters are matrices (tables) of numerical values of coefficients, and to denote the coefficients themselves use lowercase Greek letters with two indices, the first of which is the variable number, the second is the equation number, where this coefficient appears; lower-case bold letters are vectors (matrix-strings) of numerical values for the time moment indicated by the index at the vector.
Then system (1) takes the form

$$y_t = y_t(I - B) - x_t r + u_t. \quad (2)$$

Equation (2) is called the behavioral form of the simultaneous equations model.

We will proceed to the question of obtaining such an estimate of the parameters of the model of simultaneous equations, which would have better statistical properties in comparison with the estimates of both the usual least squares method and the two-step, three-step least squares method, as well as the fixed point method and other methods of estimating the coefficients of equations considered in the system simultaneously. The main drawback of methods based on minimization of the sum of squares of deviations of real observations from some theoretical dependence is that the sum of squares of errors that correlate with the values of variables included in the expression for these errors is minimized.
The random deviation, one of the elements of the right-hand side of the regression equation, depends on all of its elements: the regression coefficients, values of the dependent and "independent" variables. But the relationship between the dependent variable and the random deviation is determined by the type of the equation itself and with respect to the variables on the right side of the equation, there is one of the main postulates of the least squares method, which is that these independent variables are already predetermined, deterministic and independent of chance [4, 5].
Therefore, the error in the minimized target function is determined only by the coefficients. Then the regression coefficients are chosen so that the sum of squares of errors is the smallest. But in the case when the explained variables correlate with the error, the target function of the least squares method becomes dependent on the values of the variables themselves, then the method of least squares-estimate of regression coefficients is made conditional, that is. fair only for the given set of values of the variables. This statement is also true for multistep methods, since they assume replacement of the simultaneous equations with predetermined equations, which are not. Such a model cannot be used for the study of the economic system, because such analysis implies the study of sensitivity of the resulting indicators to all possible values of the explained quantities [6].
The logical side of the unsuitability of methods based on the principle of minimizing the sum of squares of errors was mentioned above. There is also a statistical side to this problem. Violation of the basic assumption of the least squares method of independence of predetermined variables and errors leads to the fact that the least squares method-estimates in this case lose their best properties: unbiasedness, efficiency, and consistency. Thus, the method of least squares, including the two- and three-step method, becomes a crude, very approximate method suitable only for superficial estimations, which, however, can turn out to be erroneous.

3. Results

The expectation of the product of the stochastic exogenous variables by the random deviations of the simultaneous equation is a covariance matrix between the deviations of the factor and the simultaneous equations. However $\omega_{\mathrm{Xu}_i} \neq \sigma_{\mathrm{Xu}_i}$, since ω_{Xu_i} – is the covariance of the random deviations with the stochastic exogenous variables of the i-th equation which vector has dimension n_i-*1*, where n_i – is the number of these variables, and σ_{Xu_i} – is the covariance of the random deviations of the i-th equation with all exogenous variables with dimension *n-1*.
Theorem: In the case of stochasticity of exogenous variables, the estimation of the coefficients of the simultaneous equation obtained by the method of factorization of simultaneous equations (MFSE) is more efficient than the estimation of these coefficients by the two-step least squares (2LS) method.
Proof. The covariance matrix of the 2LS-estimate has the form

$$\operatorname{cov}\left\{\begin{pmatrix}\hat{\beta}_i\\ \gamma_i\end{pmatrix}\right\} = E\left\{\left(\begin{pmatrix}\hat{\beta}_i\\ \gamma_i\end{pmatrix}_{2\mathrm{LS}} - \begin{pmatrix}\beta_i\\ \gamma_i\end{pmatrix}\right)\left(\begin{pmatrix}\hat{\beta}_i\\ \gamma_i\end{pmatrix}_{2\mathrm{LS}} - \begin{pmatrix}\beta_i\\ \gamma_i\end{pmatrix}\right)^T\right\} = [i = 1,2,\dots,k] =$$

$$= E\left\{\left(\begin{pmatrix}\hat{Y}_i^T\hat{Y}_i & \hat{Y}_i^T X_i\\ X_i^T\hat{Y}_i & X_i^T X_i\end{pmatrix}^{-1}\begin{pmatrix}\hat{Y}_i^T\\ X_i^T\end{pmatrix}y^T - \begin{pmatrix}\beta\\ \gamma\end{pmatrix}\right)\left(\begin{pmatrix}\hat{Y}_i^T\hat{Y}_i & \hat{Y}_i^T X_i\\ X_i^T\hat{Y}_i & X_i^T X_i\end{pmatrix}^{-1}\begin{pmatrix}\hat{Y}_i^T\\ X_i^T\end{pmatrix}y^T - \begin{pmatrix}\beta\\ \gamma\end{pmatrix}\right)^T\right\} =$$

$$= E\left\{\left(\begin{pmatrix} \hat{Y}_i^T\hat{Y}_i & \hat{Y}_i^T X_i \\ X_i^T\hat{Y}_i & X_i^T X_i \end{pmatrix}^{-1}\begin{pmatrix} \hat{Y}_i^T \\ X_i^T \end{pmatrix} y^T - \begin{pmatrix} \beta \\ \gamma \end{pmatrix}\right)\left(\left(A_i : X_i\begin{pmatrix} \beta \\ \gamma \end{pmatrix} + u_i\right)^{-1} - \begin{pmatrix} \beta \\ \gamma \end{pmatrix}\right)^T\right\} =$$

$$= E\left\{\left(\begin{pmatrix} \beta \\ \gamma \end{pmatrix} + \begin{pmatrix} \hat{Y}_i^T\hat{Y}_i & \hat{Y}_i^T X_i \\ X_i^T\hat{Y}_i & X_i^T X_i \end{pmatrix}^{-1}\begin{pmatrix} \hat{Y}_i^T \\ X_i^T \end{pmatrix} u_i - \begin{pmatrix} \beta \\ \gamma \end{pmatrix}\right) A^T\right\} =$$

$$= E\left\{\left(\begin{pmatrix} \hat{Y}_i^T\hat{Y}_i & \hat{Y}_i^T X_i \\ X_i^T\hat{Y}_i & X_i^T X_i \end{pmatrix}^{-1}\begin{pmatrix} \hat{Y}_i^T \\ X_i^T \end{pmatrix} u_i\right)\left(\begin{pmatrix} \hat{Y}_i^T\hat{Y}_i & \hat{Y}_i^T X_i \\ X_i^T\hat{Y}_i & X_i^T X_i \end{pmatrix}^{-1}\begin{pmatrix} \hat{Y}_i^T \\ X_i^T \end{pmatrix} u_i\right)^T\right\} =$$

$$= E\left\{\left(\begin{pmatrix} \hat{Y}_i^T\hat{Y}_i & \hat{Y}_i^T X_i \\ X_i^T\hat{Y}_i & X_i^T X_i \end{pmatrix}^{-1}\begin{pmatrix} \hat{Y}_i^T \\ X_i^T \end{pmatrix} u_i u_i^T A_i : X_i \begin{pmatrix} \hat{Y}_i^T\hat{Y}_i & \hat{Y}_i^T X_i \\ X_i^T\hat{Y}_i & X_i^T X_i \end{pmatrix}^{-1}\right)\right\} =$$

$$= E\left\{w^{-1}\begin{pmatrix} Y_i^T X(X^TX)^{-1}X^T u_i u_i^T X(X^Y X)^{-1}X^T Y_i & Y_i^T X(X^TX)^{-1}X^T u_i u_i^T X_i \\ X_i^T u_i u_i^T X(X^TX)^{-1}X^T Y_i & X_i^T u_i u_i^T X_i \end{pmatrix}\right\} \quad (3)$$

We will find the covariance matrix of the MFSE estimation of the coefficients of the simultaneous equation (FSE).

$$\text{cov}\left\{\begin{pmatrix} \hat{\beta}_i \\ \gamma_i \end{pmatrix}_{\text{FSE}}\right\} = E\left\{\left(\begin{pmatrix} \hat{\beta}_i \\ \gamma_i \end{pmatrix}_{\text{FSE}} - \begin{pmatrix} \beta_i \\ \gamma_i \end{pmatrix}\right)\left(\begin{pmatrix} \hat{\beta}_i \\ \gamma_i \end{pmatrix}_{\text{FSE}} - \begin{pmatrix} \beta_i \\ \gamma_i \end{pmatrix}\right)^T\right\} = [i = 1,2,\ldots k] =$$

$$E\left\{\left(\begin{pmatrix} A_{Y_i}A_{Y_i}^T & A_{Y_i}A_{X_i}^T \\ A_{X_i}A_{Y_i}^T & A_{X_i}A_{X_i}^T \end{pmatrix}^{-1}\begin{pmatrix} A_{Y_i} \\ A_{X_i} \end{pmatrix} F^T y_i - \begin{pmatrix} \beta_i \\ \gamma_i \end{pmatrix}\right)\left(\begin{pmatrix} A_{Y_i}A_{Y_i}^T & A_{Y_i}A_{X_i}^T \\ A_{X_i}A_{Y_i}^T & A_{X_i}A_{X_i}^T \end{pmatrix}^{-1}\begin{pmatrix} A_{Y_i} \\ A_{X_i} \end{pmatrix} F^T y_i - \begin{pmatrix} \beta_i \\ \gamma_i \end{pmatrix}\right)^T\right\} =$$

$$= E\left\{\left(\begin{pmatrix} A_{Y_i}A_{Y_i}^T & A_{Y_i}A_{X_i}^T \\ A_{X_i}A_{Y_i}^T & A_{X_i}A_{X_i}^T \end{pmatrix}^{-1}\begin{pmatrix} A_{Y_i} \\ A_{X_i} \end{pmatrix} F^T\left(\left(\text{F}A_{\text{Yi}}^T : A_{\text{Yi}}^T\begin{pmatrix} \beta \\ \gamma \end{pmatrix} + u_i\right) - \begin{pmatrix} \beta \\ \gamma \end{pmatrix}\right)\right)^T\right\} =$$

$$= E\left\{\left(\begin{pmatrix} \beta_i \\ \gamma_i \end{pmatrix} + \begin{pmatrix} A_{Y_i}A_{Y_i}^T & A_{Y_i}A_{X_i}^T \\ A_{X_i}A_{Y_i}^T & A_{X_i}A_{X_i}^T \end{pmatrix}^{-1}\begin{pmatrix} A_{Y_i} \\ A_{X_i} \end{pmatrix} F^T u_i - \begin{pmatrix} \beta \\ \gamma \end{pmatrix}\right) A^T\right\} =$$

$$= E\left\{\begin{pmatrix} A_{Y_i}A_{Y_i}^T & A_{Y_i}A_{X_i}^T \\ A_{X_i}A_{Y_i}^T & A_{X_i}A_{X_i}^T \end{pmatrix}^{-1}\begin{pmatrix} A_{Y_i} \\ A_{X_i} \end{pmatrix} F^T u_i u_i^T \text{F}A_{\text{Yi}}^T : A_{\text{Yi}}^T \begin{pmatrix} A_{Y_i}A_{Y_i}^T & A_{Y_i}A_{X_i}^T \\ A_{X_i}A_{Y_i}^T & A_{X_i}A_{X_i}^T \end{pmatrix}^{-1}\right\} =$$

$$= \sigma^2\begin{pmatrix} A_{Y_i}A_{Y_i}^T & A_{Y_i}A_{X_i}^T \\ A_{X_i}A_{Y_i}^T & A_{X_i}A_{X_i}^T \end{pmatrix}^{-1} \quad (4)$$

The properties of quadratic forms were used in the derivation [7]. We will show the positive definiteness of the matrix Q equal to the difference of the covariance matrix of the two-step least squares method (2LS) and the MFSE estimator.

$$Q = \text{cov}\left\{\begin{pmatrix} \hat{\beta}_i \\ \gamma_i \end{pmatrix}_{\text{2LS}}\right\} - \text{cov}\left\{\begin{pmatrix} \hat{\beta}_i \\ \gamma_i \end{pmatrix}_{FSE}\right\} =$$

$$= w^{-1}\begin{pmatrix} Y_i^T X(X^T X)^{-1}(\sigma^2 A_X A_X^T + \sigma_{Xu_i}\sigma_{Xu_i}^T)(X^T X)^{-1}X^T Y_i & Y_i^T X(X^T X)^{-1}\sigma_{Xu_i}\omega_{Xu_i}^T \\ \omega_{Xu_i}\sigma_{Xu_i}^T (X^T X)^{-1}X^T Y_i & \sigma^2 A_X A_X^T + \sigma_{Xu_i}\sigma_{Xu_i}^T \end{pmatrix} w^{-1} -$$

$$-\sigma^2 \begin{pmatrix} A_{Y_i}A_{Y_i}^T & A_{X_i}A_{X_i}^T \\ A_{X_i}A_{X_i}^T & A_{X_i}A_{X_i}^T \end{pmatrix}^{-1} \tag{5}$$

Since the variables of the model of simultaneous equations are known to be stochastic and interdependent, discarding the random component and expressing the variables through factor values, we obtain

$$Q =$$

$$= w^{-1}\begin{pmatrix} A_{Y_i}A_{X_i}^T(A_X A_X^T)^{-1}A_X A_{Y_1}^T(\sigma^2 A_X A_X^T + \sigma_{Xu_i}\sigma_{Xu_i}^T)(A_X A_X^T)^{-1}A_{Y_i}A_{Y_i}^T & A_{Y_i}A_{Y_i}^T(A_X A_X^T)^{-1}\sigma_{Xu_i}\omega_{Xu_i}^T \\ \omega_{Xu_i}\sigma_{Xu_i}^T(A_X A_X^T)^{-1}A_X A_{Y_1}^T & \sigma^2 A_X A_{X_i}^T + \sigma_{Xu_i}\sigma_{Xu_i}^T \end{pmatrix} w^{-1} -$$

$$-\sigma^2 \begin{pmatrix} A_{Y_i}A_{Y_i}^T & A_{Y_i}A_{X_i}^T \\ A_{X_i}A_{Y_i}^T & A_{X_i}A_{X_i}^T \end{pmatrix}^{-1} \tag{6}$$

in which case the W matrix takes the form

$$W = \begin{pmatrix} A_{Y_i}A_{X_i}^T(A_X A_X^T)^{-1}A_X A_{Y_1}^T & A_{Y_i}A_{Y_i}^T(A_X A_X^T)^{-1}A_X A_{Y_1}^T \\ A_{Y_i}A_{X_i}^T(A_X A_X^T)^{-1}A_X A_{Y_1}^T & A_{X_1}A_{X_i}^T \end{pmatrix} \tag{7}$$

Since the inverse of the covariance matrix of the MFSE estimate

$$V = \begin{pmatrix} A_{Y_i}A_{Y_i}^T & A_{Y_i}A_{X_i}^T \\ A_{X_i}A_{Y_i}^T & A_{X_1}A_{X_i}^T \end{pmatrix} \tag{8}$$

to the accuracy of the scalar factor and the idempotent matrix in its structure corresponds to the matrix W, then we will consider these matrices approximately identical $V \simeq W$. Then

$$Q = W^{-1}(S - \sigma^2 W)W^{-1}, \tag{9}$$

where

$$S = w^{-1}\begin{pmatrix} A_{Y_i}A_{X_i}^T(A_X A_X^T)^{-1}A_X A_{Y_1}^T(\sigma^2 A_X A_X^T + \sigma_{Xu_i}\sigma_{Xu_i}^T)(A_X A_X^T)^{-1}A_{Y_i}A_{Y_i}^T & A_{Y_i}A_{Y_i}^T(A_X A_X^T)^{-1}\sigma_{Xu_i}\omega_{Xu_i}^T \\ \omega_{Xu_i}\sigma_{Xu_i}^T(A_X A_X^T)^{-1}A_X A_{Y_1}^T & \sigma^2 A_X A_{X_i}^T + \sigma_{Xu_i}\sigma_{Xu_i}^T \end{pmatrix} \tag{10}$$

Since the matrix W^{-1}, as a symmetric matrix, is positively determined, it is sufficient to show the positive definiteness of the matrix difference matrices $(S - \sigma^2 W)$. Since these matrices have a block structure of the same order, let us consider the result of the difference matrices by blocks. Consider the diagonal blocks. The difference of the upper left blocks has the form

$$A_{Y_i}A_{X_i}^T(A_X A_X^T)^{-1}A_X A_{Y_1}^T\sigma_{Xu_i}\sigma_{Xu_i}^T(A_X A_X^T)^{-1}A_X A_{Y_i}^T - \sigma^2 A_X A_{X_i}^T(A_X A_X^T)^{-1}A_X A_{Y_i}^T =$$

$$= A_{Y_i}A_{X_i}^T(A_X A_X^T)^{-1}A_X A_{Y_1}^T\sigma_{Xu_i}\sigma_{Xu_i}^T(A_X A_X^T)^{-1}A_X A_{Y_i}^T +$$

$$+A_{Y_i}A_{X_i}^T(A_X A_X^T)^{-1}A_X A_{Y_1}^T\sigma_{Xu_i}\sigma_{Xu_i}^T(A_X A_X^T)^{-1}A_X A_{Y_i}^T - \sigma^2 A_X A_{X_i}^T(A_X A_X^T)^{-1}A_X A_{Y_i}^T =$$

$$= A_{Y_i}A_{X_i}^T(A_X A_X^T)^{-1}A_X A_{Y_1}^T\sigma_{Xu_i}\sigma_{Xu_i}^T(A_X A_X^T)^{-1}A_X A_{Y_i}^T, \tag{11}$$

and since it is a symmetric matrix, it is positively defined.

The difference of the lower right blocks is equal to $\sigma^2 A_X A_{X_i}^T + \sigma_{\mathrm{Xu}_i}\sigma_{\mathrm{Xu}_i}^T - \sigma^2 A_X A_{X_i}^T = \sigma_{X_i u_i}\sigma_{\mathrm{Xu}_i}^T\sigma_{X_i u_i}^T$, and is also a symmetric positively definite matrix. Since the off-diagonal blocks of matrix S are symmetric and decrease by the same symmetric value, the result of matrix difference $(S - \sigma^2 W)$ is a symmetric positively determined matrix. Hence,

$$Q = \mathrm{cov}\left\{\begin{pmatrix}\beta_i \\ \gamma_i\end{pmatrix}_{2\mathrm{LS}}\right\} - \mathrm{cov}\left\{\begin{pmatrix}\beta_i \\ \gamma_i\end{pmatrix}_{\mathrm{FSE}}\right\} > 0; \Rightarrow \mathrm{cov}\left\{\begin{pmatrix}\beta_i \\ \gamma_i\end{pmatrix}_{2\mathrm{LS}}\right\} > \mathrm{cov}\left\{\begin{pmatrix}\beta_i \\ \gamma_i\end{pmatrix}_{\mathrm{FSE}}\right\}, (\forall i) \quad (12)$$

the estimation of the equation coefficients obtained using the method of factorization of simultaneous equations is effective with respect to the estimation of the two-step method of least squares.

4. Discussion and conclusions

Similarly, we can show the greater efficiency of the proposed method compared to the three-step least squares method, the second most frequently used method. Thus, in the case of interdependence of equation variables of the system it is appropriate to use the method of factorization of simultaneous equations. The study of the nature of simultaneity of variables in a system of econometric equations has shown that not only endogenous variables but also exogenous variables can be interrelated in such a system. All existing by now methods of estimation of parameters of the system of simultaneous equations ignore this fact which leads to bias of estimates obtained with the help of these methods [8, 9]. The method of factorization of the system of simultaneous equations (MFSE) has been developed, which gives unbiased estimates of its parameters and these estimates are more effective compared to the most widely used methods of estimating the coefficients of the equations of the system. The MFSE has expanded the multi-model approach to predicting nonlinear non-stationary processes in optimal control problems [10].

References

[1] Kanatnikov A N 2018 New criteria of equilibrium points stability for discrete-time systems *14th International Conference Stability and Oscillations of Nonlinear Control Systems (Proceedings of 2018 Pyatnitskiys Conference)* (Moscow: STAB) pp 1–3

[2] Kravets O Ja, Shaytura S V, Minitaeva A M and Atlasov I V 2020 Analysis of routing processes in telecommunication networks with unsteady flows using markov processes *IOP Conference Series: Materials Science and Engineering* **862** (Krasnoyarsk: Krasnoyarsk Science and Technology City Hall of the Russian Union of Scientific and Engineering Associations) p 52005

[3] Minitaeva A M 2022 Analysis of the multi-criteria decision-making problem under conditions of heterogeneous interval uncertainty *4th International Conference on Control Systems, Mathematical Modeling, Automation and Energy Efficiency* (Lipetsk: SUMMA) pp 42-45

[4] Zimin V N, Krylov A V, Kuvyrkin G N and Shakhverdov A O 2022 Development of a mathematical model of a force actuator for the opening of a space structure with transformable configuration *Journal of Physics: Conference Series* **2231**

[5] Kanatnikov A N 2019 Localizing Sets and Behavior of Trajectories of Time *Varying Systems - Differential Equations* **55** pp 1420–30

[6] Minitaeva A M 2022 Decision making under interval setting of decision makers' preferences *Information Technologies in Management (Proceedings of the conference ITM-2022)* (St. Petersburg: ETU "LETI") pp 198-201

[7] Kotz S, Read C B, Balakrishnan N and Vidakovic B 2005 *Encyclopedia of Statistical Sciences* (New York: John Wiley & Sons)

[8] Suyatinov S I, Buldakova T I and Vishnevskaya Yu A 2021 Identification of Situations based on Synergetic Model *3rd International Conference on Control Systems, Mathematical Modeling, Automation and Energy Efficiency* (Lipetsk: SUMMA) pp 509-14

[9] Syuzev V V, Smirnova E V and Proletarsky A V 2021 Algorithms of multidimensional random process simulation *Computer Optics* **45** pp 627-37

[10] Minitaeva A M 2022 Multimodel approach to predicting nonlinear nonstationary processes in optimal control problems *International Conference "Applied Mathematics, Computational Science and Mechanics: Current Problems"* (Voronezh: VSU) p 1865

Critical temperature in 2D square Ising model with anisotropic coupling constant using Kramers-Wannier duality

L A Aviles, R P Gammag*

Department of Physics, Mapúa University, Manila, Philippines

* Corresponding author's e-mail: rpgammag@mapua.edu.ph

Abstract. To solve the 2D square Ising model incorporating anisotropy, we discuss a graphical representation of Kramers-Wannier duality. The method yields an equation relating the critical temperature (T_c) to the horizontal and vertical coupling constants. The result reproduces the known exact solution by Onsager The equation derived was utilized to plot T_c as a function of anisotropy by introducing the ratio (α) of the coupling constants. A phase diagram is then obtained showing T_c as an increasing function of α. In the limit as α approaches zero, the critical temperature asymptotically goes to zero. This agrees with the exact solution of the 1D model.

1. Introduction

The Ising model is a fair assumption in predicting the macroscopic behaviour of a system involving spins in a lattice. It is mostly used in studying the phase transition of magnetic materials despite its oversimplification of magnets. As mentioned by Wolf [1], there are very clear differences between the theory and actual magnetic materials. However, experiments conducted on 2D materials [2] such as K_2CoF_4 show good comparison with theoretical predictions and a few adjustments to the exact solution of the 2D model to fit experimental results with Rb_2CoF_4.

The exact solution for the 1D model can easily be obtained since the work of Ising [3] yet practical implications made it difficult for experimental application. Solutions for the 2D model are also relatively obtainable and gained better applications since the emergence of 2D magnetic materials. One of the most recognized and earliest works on the 2D model is the exact solution done by Onsager. In his work [4], he considered a 2D square lattice and obtained the critical temperature T_c using the following,

$$\sinh\left(\frac{2J}{k_B T_c}\right)\sinh\left(\frac{2J'}{k_B T_c}\right) = 1, \quad (1)$$

where k_B is the Boltzmann constant. This equation yields the phase transition point (T_c) for a certain value of the horizontal (J) and vertical (J') coupling constant.

Prior to Onsager's solution, equation (1) is already established by Kramers and Wannier. However, their solution [5] lacks some of the properties to completely describe a magnetic system. Regardless, Onsager mentioned the importance of their work in his exact solution to determine T_c. Subsequently, the analytical results from [4] and [5], particularly equation (1), became the point of reference for every method in solving the 2D Ising model and are well regarded as exact solutions.

In this paper, we also consider a 2D square Ising model where two coupling constants link the neighbouring spins in a nearest-neighbour chain (NN interactions). In the following sections, we will discuss how to solve for equation (1) by formulating a set of rules following the graphical representation of the Kramers-Wannier duality made by Baxter [6]. Using equation (1) and the ratio

$\alpha = \frac{J'}{J}$, we investigate the anisotropic region, where $\alpha = 1$ is the isotropic case. By plotting T_c as a function α, we show a wider range of T_c for different anisotropies. The results show similar behaviour of T_c compared with solutions that incorporate mean-field approximation (MFA) [7,8] and numerical analysis from Monte Carlo simulations (MCS) [9].

2. Overview and assumptions of the Ising Model

The Ising model was first introduced by Lenz [10]. It assumes a system composed of spins fixed at a lattice structure. The orientation of each spin σ is either up (+) or down (-). The interaction between neighbouring spins depends on the temperature of the system. Below T_c, most of the spins are parallel and sustains a net magnetic moment thus maintaining an ordered phase. Above T_c, the system is in a disordered phase where the spins are randomly oriented losing the overall magnetic moment.

The problem to derive a closed form of the partition function Z_N. Using Z_N, one can derive all the thermodynamic functions that describes the system's macroscopic state (e.g. phase) from its microscopic properties (e.g. spins). One case is using Z_N to obtain the free energy F. If there is a discontinuity in F, the system undergoes a phase transition.

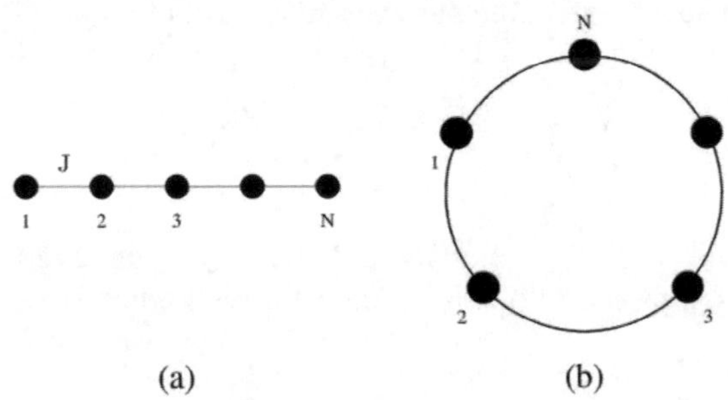

Figure 1. (a) 1D Ising model, (b) 1D Ising model with a periodic boundary condition.

Ising obtained the free energy for an isotropic 1D lattice Ising model and found that there is no phase transition at any finite temperature. The system is also isotropic, indicating that the magnitude of the coupling constant J is equal throughout the neighbouring nodes. To determine F, a periodic boundary condition is assumed so that the system can be considered finite despite the number of sites (N) approaching infinity (Figure 1). This condition is useful even at higher dimensions as it helps with the calculations such that there are no source terms at the endpoints and Z_N can be written in closed form.

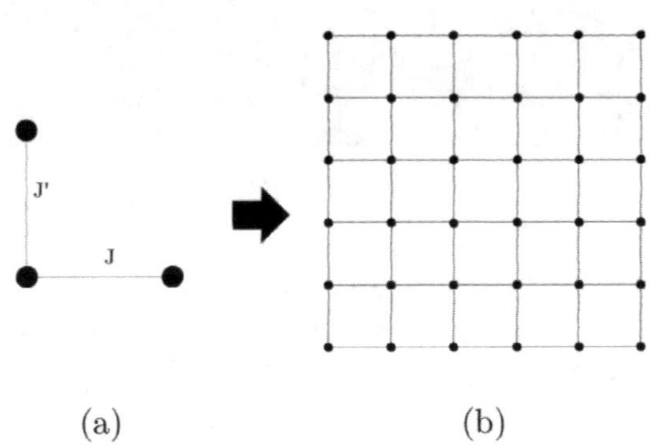

Figure 2. (a) 2D unit cell, (b) 2D unit cell forming a primitive lattice.

In a 2D square lattice, a spin interacts with two noncolinear neighbouring spins to form a unit cell

Journal of Physics: Conference Series **2543** (2023) 012006 doi:10.1088/1742-6596/2543/1/012006

(Figure 2). Onsager presented an exact and analytical solution of the free energy for a 2D Ising model. The resulting free energy is discontinuous at a point similar to the expression obtained in Kramers-Wannier duality (KWD) which is given by equation (1). Unfortunately, KWD does not provide a closed form of Z_N hence it cannot be used in calculating thermodynamic functions.

3. Kramers-Wannier duality of a 2D square Ising model

We will solve the 2D square Ising model using KWD and obtain T_c without going through the free energy. This is done by comparing the summation of the low and high-temperature expansion of Z_N. As shown by Baxter [6], the process is made easier by making a graphical correspondence with each term in the summation.

To obtain the partition function, we start by writing the Hamiltonian of the system given by,

$$\mathcal{H} = -J\sum_{(i,j)} \sigma_i\sigma_j - J'\sum_{(i,k)} \sigma_i\sigma_k . \tag{2}$$

In equation (2), $\sigma = \pm 1$, representing the orientation of the spins. The first sum is the energy contribution from the horizontal edges (i, j) and the second is from the vertical edges (i, k). Here, the word "edges" refers to the unique pair of spins in a nearest neighbour chain.

Using the total energy in equation (2), the partition function can now be written as follows,

$$Z_N = \sum_{\sigma} \exp\left[K\sum_{(i,j)} \sigma_i\sigma_j + L\sum_{(i,k)} \sigma_i\sigma_k\right], \tag{3}$$

where $K = J/k_BT$ and $L = J'/k_BT$. The outer sum represents the sum over all possible configurations.

As an aid in the analysis, a second lattice is introduced (see Figure 3).

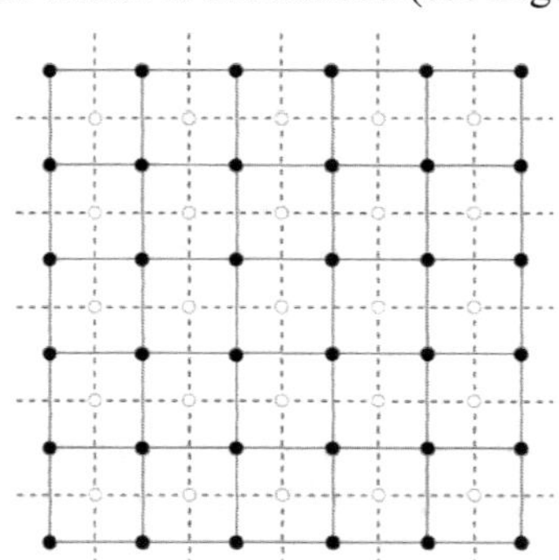

Figure 3. Original lattice $\mathcal{L}$ (solid dots) and its dual lattice $\mathcal{L}_\mathcal{D}$ (hollow dots).

In Figure 3, the original lattice $\mathcal{L}$ is where the spins are originally structured. The dual lattice is obtained by placing sites of the original lattice $\mathcal{L}$ at the centre of the dual lattice $\mathcal{L}_\mathcal{D}$.

To locate T_c, a graphical correspondence is made on $\mathcal{L}$ and $\mathcal{L}_\mathcal{D}$ for the high- and low-temperature expansion of Z_N.

3.1. Graphical correspondence for the low-temperature limit

On the original lattice $\mathcal{L}$, let M be the number of edges in the vertical and horizontal direction. For a given set of configurations, let r and s be the number of antiparallel vertical and horizontal links, respectively. Thus, the *Boltzmann weight* from equation (3) can now be written as,

$$\exp[K(M - 2s) + L(M - 2r)]. \tag{4}$$

Notice how the Boltzmann weight only depends on the antiparallel links in the system. Consequently, the spins from original lattice $\mathcal{L}$ can now be graphically defined using the dual lattice $\mathcal{L}_\mathcal{D}$ by applying the following set of rules:

(i) For every pair of antiparallel spins in the horizontal direction, a vertical line is drawn along the edge of $\mathcal{L}_{\mathcal{D}}$ crossing the spins.
(ii) For every pair of antiparallel spins in the vertical direction, a horizontal line is drawn along the edge of $\mathcal{L}_{\mathcal{D}}$ crossing the spins.
(iii) If the spins are parallel, no line is drawn.

Therefore, r and s are now accordingly defined as the number of horizontal and vertical lines drawn on $\mathcal{L}_{\mathcal{D}}$. The integer value of r and s is always even since there is always an even number of lines as seen in Figure 4. Consequently, these lines are joined to form a polygon on the dual lattice $\mathcal{L}_{\mathcal{D}}$. These polygons divide the system into up-spin or down-spin domains (see Figure 4).

The partition function in (3) can now be expressed using the Boltzmann weight from equation (4),

$$Z_N = 2\sum_{\tilde{P}} \exp[K(M-2s) + L(M-2r)] = 2\exp[M(K+L)]\sum_{\tilde{P}} \exp[-2(Lr+Ks)]\,. \quad (5)$$

Here, the summation over $\tilde{P}$ denotes the number of lines on every possible set of polygons drawn on dual lattice $\mathcal{L}_{\mathcal{D}}$. For a given set of polygons formed, another possible configuration is obtained by negating the current spins hence the factor 2 outside the sum.

It is also important to note that the expression from equation (5) is also true in the high temperature limit. However, it is convenient for the series expansion of the low-temperature limit since the first assumption is that all the spins are aligned ($r = 0, s = 0$). The perturbation is done by gradually adding more antiparallel domains for each term in the summation.

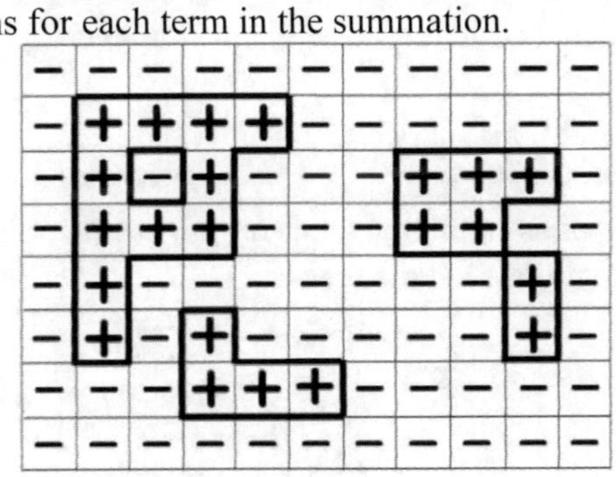

Figure 4. The dual lattice $\mathcal{L}_{\mathcal{D}}$ where domains of up-spins are separated by lines that forms a polygon. Thick lines are drawn along the edge of $\mathcal{L}_{\mathcal{D}}$ for every pair of antiparallel spins.

3.2. Graphical correspondence for the high-temperature limit

In the high-temperature limit the system is in a disordered phase. The perturbation does not depend on the direction of the spin but rather on each interacting pair of spins that are considered.

Since σ_i and σ_j can only be ± 1, we can use the identity,

$$\exp(x\sigma_i\sigma_j) = \cosh(x)\,(1 + \sigma_i\sigma_j\tanh(x)). \quad (6)$$

The partition function in equation (3) can now be written as

$$Z_N = \sum_{\sigma}\left[\prod_{(i,j)}\cosh(K)\,(1+\sigma_i\sigma_j\tanh(K))\prod_{(i,k)}\cosh(L)\,(1+\sigma_i\sigma_k\tanh(L))\right], \quad (7)$$

$$Z_N = [\cosh(K)\cosh(L)]^M\sum_{\sigma}\left[\prod_{(i,j)}(1+v\sigma_i\sigma_j)\prod_{(i,k)}(1+w\sigma_i\sigma_k)\right], \quad (8)$$

where $v = \tanh(K)$ and $w = \tanh(L)$. The first product is over all M horizontal edges (i,j) and the second product is over all M vertical edges (i,k).

The process of expanding the summation is made easier by making a one-to-one correspondence between the terms in the expansion and a graphical representation on the original lattice $\mathcal{L}$:

(i) For every factor of $v\sigma_i\sigma_j$ (interacting spins in the horizontal direction), a horizontal line is drawn along the edge of $\mathcal{L}$ connecting the spins.

(ii) For every factor of $w\sigma_i\sigma_k$ (interacting spins in the vertical direction), a vertical line is drawn along the edge of $\mathcal{L}$ connecting the spins.

(iii) Rules 1 and 2 are only valid if for every factor of $v\sigma_i\sigma_j$ and $w\sigma_i\sigma_k$, the spins σ_i, σ_j, and σ_k interact with even number of neighbours.

The third rule is added to avoid cancelling terms of $v\sigma_i\sigma_j$ and $w\sigma_i\sigma_k$ when summed over all the possible configurations. Consequently, the implementation of the third rule ensures that there is always an even number of lines drawn eventually joining the lines to form a polygon.

Using the following set of rules, the product in equation (8) takes the form of

$$v^r w^s \sigma_1^{n_1} \sigma_2^{n_2} \sigma_3^{n_3} \ldots, \tag{9}$$

where r and s are the number of horizontal and vertical lines drawn, respectively. The variable n_i is the number of sites σ_i as an endpoint. Since n_i is always even and $\sigma = \pm 1$, the factor $\sigma_1^{n_1}\sigma_2^{n_2}\sigma_3^{n_3}$ ... is always positive. When summed for all possible configurations a factor of 2^N will be formed, therefore the partition function from equation (8) can be written using the expression from equation (9),

$$Z_N = 2^N [\cosh(K)\cosh(L)]^M \sum_P v^r w^s. \tag{10}$$

Here, the summation over P denotes the number of lines on every possible set of polygons drawn on the original lattice $\mathcal{L}$.

3.3. *Duality between the expansions*

The sum for both the series expansion only depends on the line drawn over the closed polygon P and $\tilde{P}$ which is evident in both equation (5) and equation (10),

$$\sum_{\tilde{P}} \exp[-2(Lr + Ks)], \tag{11}$$

$$\sum_P v^r w^s. \tag{12}$$

Duality is observed between equations (11) and (12) which are related by the exponents s and r,

$$v = e^{-2\tilde{L}}; \tag{13a}$$

$$w = e^{-2\tilde{K}}, \tag{13b}$$

where $\tilde{L}$ and $\tilde{K}$ denote the low-temperature limit. Inserting $v = \tanh(K)$ and $w = \tanh(L)$ to equation (13a) and (13b),

$$\tanh(K) = e^{-2\tilde{L}}, \tag{14a}$$

$$\tanh(L) = e^{-2\tilde{K}}. \tag{14b}$$

Using the properties of hyperbolic functions to rewrite equation (14a),

$$\frac{\sinh(2K)}{\cosh(2K)+1} = \frac{1-e^{-4\tilde{L}}}{2\sinh(2\tilde{L})}, \tag{15}$$

$$\sinh(2K)\sinh(2\tilde{L}) = \frac{1}{2}(\cosh(2K)+1)(1-e^{-4\tilde{L}}). \tag{16}$$

Performing the same method for equation (14b) we get the same form from equation (16).

We then apply a mathematical limit at the right-hand side of equation (16) where at low temperature $\tilde{L}$ and $\tilde{K}$ are large while at high temperature L and K are small, we have,

$$\sinh(2K)\sinh(2\tilde{L}) = 1, \tag{17a}$$

$$\sinh(2\tilde{K})\sinh(2L) = 1. \tag{17b}$$

With only one line of critical temperature in the (K, L) plane, the temperature limits must converge at $K = \widetilde{K} = K_c$ and $L = \widetilde{L} = L_c$. Both equation (17a) and equation (17b) can now be written as,

$$\sinh(2K_c)\sinh(2L_c) = 1. \tag{18}$$

This was verified by Onsager in which the free energy is discontinuous at $\sinh(2K)\sinh(2L) = 1$. Since $K_c = J/k_BT_c$ and $L_c = J'/k_BT_c$,

$$\sinh\left(\frac{2J}{k_BT_c}\right)\sinh\left(\frac{2J'}{k_BT_c}\right) = 1. \tag{19}$$

In the isotropic case where $J = J'$, the critical point is around $k_BT_c/J = 2/\ln(1+\sqrt{2}) \simeq 2.269185$.

4. Anisotropy between J and J'

Using equation (19), we check the effects of anisotropy on T_c. We will only consider anisotropy between J and J'. As mentioned, it is assumed that there is one line of critical temperature in the (K, L) plane. Hence, a specific value of α corresponds to a single critical temperature k_BT_c/J.

Table 1. Effects of anisotropy to the critical temperature between MCS from [9] and KWD.

$\alpha = \frac{J'}{J}$	k_BT_c/J (MCS)	k_BT_c/J (KWD)
0.25	1.235	1.239
0.50	1.638	1.641
1.00 (Isotropic case)	2.264	2.269
1.50	2.801	2.801
2.00	3.276	3.283

We can verify the critical temperatures obtained by MCS done by Farsal, et al. [9], in which they simulate a square lattice of size $N = S \times S$ for $S = \{8,16,32,64\}$ and rely on finite-size scaling analysis using a *Binder cumulant*. This is done by comparing the Binder parameter for each finite size of N to obtain a critical temperature close to the exact solution where the size of N is infinite (see Table 1). It is difficult to generate data points for a continuous α hence the incremental values given in table 1. It is why some MCS, such as the one by Preis, et al. [11], demonstrated the use of better hardware, thereby significantly reducing simulation times.

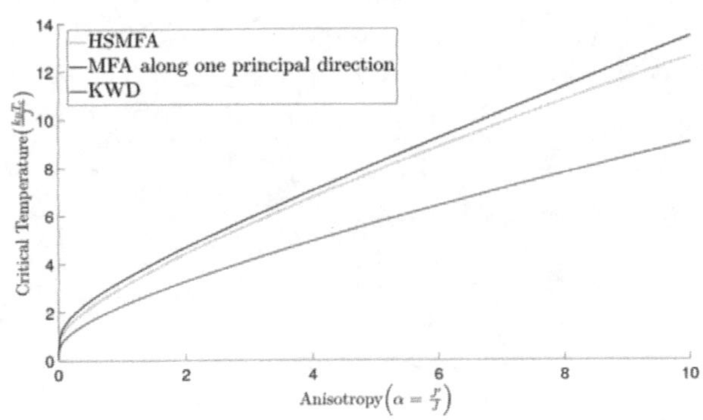

Figure 5. The plot of the rescaled critical temperature k_BT_c/J as a function of $\alpha = J'/J$ from three different solutions. Here, we observe solutions that incorporate MFA along one principal direction [7] and hard-spin mean field approximation (HSMFA) [8] compared with the exact solution from KWD.

In contrast with table 1, here in figure 5 we show a continuous value of α with k_BT_c/J using equation (19). Below the curve is the ordered phase where most of the spins are parallel. As T approaches T_c, the spins become more randomized eventually reaching the disordered phase where half of the spins points in one direction and the other half in the opposite. Furthermore, the relationship between T_c and α suggests that T_c can be written as a function of anisotropy (i.e. $k_B T_c/J = f(\alpha)$).

We also compare the behaviour of the curve to the solutions obtained by incorporating MFA along one principal direction [7] and HSMFA [8]. The mean-field theory ignores the interaction between neighbouring edges but instead considers the average behaviour of the spins. Here, we observe the curves sharing a similar behaviour with the exact solution from KWD but exceeding the critical temperature as we go up in α. An excellent comparison is observed in [12] in which they compiled solutions that are not mentioned here in plotting T_c as a function of α.

5. Summary and conclusion

The Ising model is one of the most famous, if not, the most popular model in statistical physics dealing with phase transitions. The graphical approach of Kramers-Wannier duality made it simpler to obtain T_c compared to the method made by Onsager. From the phase diagram obtained from KWD, we see the difference between an Ising model with anisotropic and isotropic interactions.

In the limit as α approaches zero, T_c asymptotically goes to zero, in agreement with the exact solution of the 1D model. For higher anisotropies ($\alpha > 1$), the critical temperature increases continuously and almost behaves in a linear manner. The same trend can also be seen in the mean-field approach. The Monte Carlo simulation has a much closer value of T_c for a given α which is predicted when compared to the exact solution.

References

[1] Wolf W 2000 *Braz. J. Phys.* **30** 794–810
[2] Breed D, Gilijamse K and Miedema A 1969 *Physica* **45** 205–216
[3] Ising E 1925 *Z. Physik* **31**(1) 253–258
[4] Onsager L 1944 *Phys. Rev.* **65**(3-4) 117–149
[5] Kramers H A and Wannier G H 1941 *Phys. Rev.* **60**(3) 252–262
[6] Baxter R 1985 Exactly solved models in statistical mechanics *Integrable Systems in Statistical Mechanics* (World Scientific) pp 5–63
[7] Garrido P and Marro J 1987 *J. Phys. A: Math. Gen.* **20** 1829–1838
[8] Çaglar T and Berker A N 2011 *Phys. Rev. E* **84**(5) 051129
[9] Farsal D, Snina M, Badia M and Bennai M 2017 *J. Supercond. Nov. Magn.* **30** 2187–2195
[10] Brush S G 1967 *Rev. Mod. Phys.* **39** 883
[11] Preis T, Virnau P, Paul W and Schneider J J 2009 *J. Comput. Phys.* **228** 4468–4477
[12] Neto M A, dos Anjos R A and de Sousa J R 2006 *Phys. Rev. B* **73**(21) 214439

Analytical solutions to the isothermal Euler equations with time-dependent damping

Xinyue Jia

Department of Mathematics, Northwest University, Xi'an, 710069, China

Corresponding author's e-mail: 202020662@stumail.nwu.edu.cn

Abstract. In this paper, we construct a class of analytical solutions to the one dimensional compressible isothermal Euler equations with time-dependent damping. By introducing a special density function $\rho(x,t)=e^{c(t)x+d(t)}$, we obtain a family of analytical solutions. A sufficient condition for the solution to blow up in finite time is given. On the basis of the conclusion, the analytical solutions to the initial boundary value problem of the pressureless Euler equations are obtained.

1. Introduction

we consider the compressible isothermal Euler equations with time-dependent damping in one dimension

$$\begin{cases} \rho_t+(\rho u)_x=0,(x,t)\in\mathbb{R}\times\mathbb{R}^+, \\ (\rho u)_t+(\rho u^2)_x+[a\rho]_x=-\dfrac{\mu}{(1+t)^\lambda}\rho u,(x,t)\in\mathbb{R}\times\mathbb{R}^+, \end{cases} \tag{1.1}$$

where, $\rho(x,t)$, $u(x,t)$ stands for the density and the velocity, and $a\rho$ with $a>0$ is the pressure of the flow. This means the flow is the isothermal. The term $-\dfrac{\mu}{(1+t)^\lambda}\rho u$ with $\mu>0$, $\lambda\geqslant 0$ in the system (1.1) is the so-called damping effect on the fluid.

There is huge literature on the studies of the global existence to the 1D compressible Euler system with time-dependent damping. In the case with $\mu=0$, the system (1.1) is the classical Euler equations which has been extensively studied. Courant et al.[1]and Li et al.[2] pointed that the solutions for Euler equations blow up due to the formation of shocks (see also[3,4]). In the case with $\lambda=0$ and $\mu>0$, the existence of global classical solution for the initial boundary value problem of the compressible Euler equations with time-dependent damping with the initial data are small perturbations near constant states for initial density away from vacuum has been proved by Hsiao[5] (see also [6]). However, when the vacuum is allowed initially, the situation is different. In [7], Liu converted the Cauchy problem of (1.1) into the initial boundary value problem of p-system on a bounded interval, and obtained the existence of local solutions when the sound velocity on the contact surface of fluid and vacuum meets certain conditions. In the case with $\lambda\neq 0$ and $\mu>0$, Pan[8] has proved that the existence of classical solutions of Euler equations, if initial data are small and compact perturbations of constant states for the case $0<\lambda\leqslant 1$ and $\mu>0$ or $\lambda=1$ and $\mu>2$. For $\lambda>1$ and

$\mu > 0$ or $\lambda = 1$ and $0 \leqslant \mu \leqslant 2$, Pan[9,10] has proved that solutions of Euler equations can blow up under some conditions on initial data. Later, Sugiyama et al.[11,12], Dong et al.[13] and Chen et al.[14] improved Pan's work. In [15], the authors constructed some radial symmetric solutions with $\rho_r = 0$ or $u = c(t)r$ to the 1D compressible Euler equations with time-depending damping. In this paper, we are interested in the construction of non-radially symmetric solutions to compressible isothermal Euler equations with time-depending damping.

2. Main result

We are now in position to state our main theorem.

Theorem 2.1 There exists an analytical solution to (1.1) with the form

$$\rho(x,t) = e^{c(t)x+d(t)}, u = -\frac{c'(t)}{c(t)}x + \frac{c'(t) - c(t)d'(t)}{c(t)^2}, \tag{2.1}$$

Where

(i) if $\lambda = 1$ and $\mu = 1$, then

$$c(t) = \frac{1}{z_0 + z_1 \log(1+t)},$$

$$d(t) = d_0 + \frac{d_1}{z_0 + z_1 \log(1+t)} - \log|z_0 + z_1 \log(1+t)| + \int_0^t \frac{a(1+s)}{2[z_0 + z_1 \log(1+s)]^2} ds,$$

where d_0, d_1 are constants. In particular, if $z_0 z_1 < 0$, the analytical solution (2.1) blow up at time $t_0 = e^{-z_0 z_1^{-1}} - 1$, otherwise the above solution is a global solution;

(ii) if $\lambda = 1$ and $\mu \neq 1$, then

$$c(t) = \frac{1-\mu}{z_0(1-\mu) + z_1[(1+t)^{1-\mu} - 1]},$$

$$d(t) = d_0 + \frac{d_1}{z_0(1-\mu) + z_1[(1+t)^{1-\mu} - 1]} - \log\left|z_0 + \frac{z_1}{1-\mu}[(1+t)^{1-\mu} - 1]\right| + \int_0^t \frac{a(1-\mu)^2(1+s)}{(\mu+1)\left[z_0 + \frac{z_1}{1-\mu}[(1+s)^{1-\mu} - 1]\right]^2} ds,$$

where d_0, d_1 are constants. In particular, if $0 < \mu < 1$, $z_0 z_1 < 0$ or $\mu > 1$, $-1 < z_1^{-1} z_0(\mu - 1) < 0$, then the solutions (2.1) blow up at time $t_0 = [1 + z_1^{-1} z_0(\mu-1)]^{(1-\mu)^{-1}} - 1$, otherwise the above solution is a global solution;

(iii) if $\lambda \neq 1$, then

$$c(t) = \frac{1}{z_0 + z_1 \int_0^t e^{\frac{\mu}{\lambda-1}[(1+s)^{1-\lambda} - 1]} ds},$$

$$d(t) = d_0 + \int_0^t c^2(s) e^{\frac{\mu}{\lambda-1}[(1+s)^{1-\lambda} - 1]} \left[d_1 - \frac{z_1}{c(s)} + \int_0^s a e^{\frac{\mu}{1-\lambda}[(1+\xi)^{1-\lambda} - 1]} d\xi \right] ds,$$

where d_0, d_1 are constants. In particular, if $0 \leqslant \lambda < 1$, $z_0 z_1 < 0$ and $1 + z_0^{-1} z_1 \int_0^{+\infty} e^{\frac{\mu}{\lambda-1}[(1+s)^{1-\lambda} - 1]} ds < 0$ or $\lambda > 1$, $z_0 z_1 < 0$, then the solutions (2.1) blow up at time t_0, which satisfies $z_0 + z_1 \int_0^{t_0} e^{\frac{\mu}{\lambda-1}[(1+s)^{1-\lambda} - 1]} ds = 0$, otherwise the above solution is a global solution.

3. Proof of Theorem 2.1

First, it is easy to verify that (2.1) satisfies the mass equation $(1.1)_1$ by direct calculation. Then, the equation $(1.1)_2$ gives

$$u_t + uu_x + a\frac{\rho_x}{\rho} + \frac{\mu}{(1+t)^\lambda}u = 0.$$

We plug (2.1) into the above equation to yields

$$\begin{aligned}
&u_t + uu_x + a\frac{\rho_x}{\rho} + \frac{\mu}{(1+t)^\lambda}u \\
&= [\frac{2c'(t)^2}{c(t)^2} - \frac{c''(t)}{c(t)} - \frac{\mu c'(t)c(t)}{(1+t)^\lambda}]x + ac(t) \\
&+ \frac{c(t)c''(t) - c(t)^2 d''(t) - 3c'(t)^2 + 2c(t)c'(t)d'(t)}{c^3(t)} + \frac{\mu c(t)c'(t) - \mu c(t)d'(t)}{(1+t)^\lambda c(t)^3} = 0.
\end{aligned}$$

For the above polynomials, a system of differential equations with respect to $(c(t), d(t))$ can be obtained by using the comparison coefficient method

$$\begin{cases}
2c'(t)^2 - c(t)c''(\mathrm{t}) - \frac{\mu c(\mathrm{t})c'(\mathrm{t})}{(1+\mathrm{t})^\lambda} = 0, \\
d''(\mathrm{t}) - [2\frac{c'(\mathrm{t})}{c(\mathrm{t})} - \frac{\mu}{(1+\mathrm{t})^\lambda}]d'(\mathrm{t}) = ac(\mathrm{t})^2 - \frac{c'(\mathrm{t})^2}{c(\mathrm{t})^2}.
\end{cases} \tag{3.1}$$

We start with equation $(2.2)_1$ to solve the function $c(t)$. Define a new function $z(t)$. which satisfies

$$z(t) = c^{-1}(t)$$

with $z(0) \neq 0$. It then follows from (2.2) that

$$\begin{cases}
z''(t) + \frac{\mu}{(1+t)^\lambda}z'(t) = 0, \\
d''(t) + [2\frac{z'(t)}{z(t)} + \frac{\mu}{(1+t)^\lambda}]d'(t) = \frac{a - z'(t)^2}{z(t)^2}.
\end{cases} \tag{3.2}$$

Through the above ordinary differential equation and the monotone analysis of the function, we deduce that

(i) if $\lambda = 1$ and $\mu = 1$, then

$$z(t) = z_0 + z_1 \log(1+t),$$
$$c(t) = [z_0 + z_1 \log(1+t)]^{-1}.$$

If $z_0 z_1 < 0$, then $\lim_{t \to t_0^-} z(t) = 0$, where $t_0 = e^{-z_0 z_1^{-1}} - 1$. According to the existence theorem of ordinary differential equation, $d'(t)$ must exist on $t \in [0, t_0)$ and

$$d'(t) = \frac{k + a(1+t) - 2z_1^2 \log(1+t)}{2[z_0 + z_1 \log(1+t)]^2},$$

$$d(t) = d_0 + \frac{d_1}{z_0 + z_1 \log(1+t)} - \log|z_0 + z_1 \log(1+t)| + \int_0^t \frac{a(1+s)}{2[z_0 + z_1 \log(1+s)]^2} ds,$$

where, k, d_0, d_1 are constants. It is not difficult to find that there is $x_0 \in \mathbb{R}$, such that

$$\lim_{t \to t_0^-} \rho(x_0, t) = +\infty.$$

Otherwise, the above $(c(t),d(t))$ is the global solution of the system of differential equations (3.1), so the analytical solution (2.1) is the global solution of the system (1.1);
(ii) if $\lambda=1$ and $\mu\neq 1$, then

$$z(t)=z_0+\frac{z_1}{1-\mu}[(1+t)^{1-\mu}-1],$$

$$c(t)=\frac{1-\mu}{z_0(1-\mu)+z_1[(1+t)^{1-\mu}-1]}.$$

If $0<\mu<1$, $z_0z_1<0$ or $\mu>1$, $-1<z_1^{-1}z_0(\mu-1)<0$, then there exists t_0 which satisfies $t_0=[1+z_1^{-1}z_0(\mu-1)]^{(1-\mu)^{-1}}-1$ such that $\lim\limits_{t\to t_0^-}z(t)=0$. According to the existence theorem of ordinary differential equation, $d'(t)$ must exist on $t\in[0,t_0)$ and

$$d'(t)=\frac{k+a(1+t)^{\mu+1}+(\mu+1)(\mu-1)z_1^{\ 2}(1+t)^{1-\mu}}{(\mu+1)(1+t)^{\mu}[z_0+z_1(1+t)^{1-\mu}]^2},$$

$$d(t)=d_0+\frac{d_1}{z_0(1-\mu)+z_1[(1+t)^{1-\mu}-1]}-\log|z_0+\frac{z_1}{1-\mu}[(1+t)^{1-\mu}-1]|+\int_0^t\frac{a(1-\mu)^2(1+s)}{(\mu+1)\left[z_0+\frac{z_1}{1-\mu}[(1+s)^{1-\mu}-1]\right]^2}\mathrm{d}s,$$

where, k,d_0,d_1 are constants. It is not difficult to find that there is $x_0\in\mathbb{R}$, such that

$$\lim_{t\to t_0^-}\rho(x_0,t)=+\infty.$$

Otherwise, the above $(c(t),d(t))$ is the global solution of the system of differential equations (3.1), so the analytical solution (2.1) is the global solution of the system (1.1);
(iii) if $\lambda\neq 1$, then

$$z(t)=z_0+z_1\int_0^t e^{\frac{\mu}{\lambda-1}[(1+s)^{1-\lambda}-1]}\mathrm{d}s,$$

$$c(t)=[z_0+z_1\int_0^t e^{\frac{\mu}{\lambda-1}[(1+s)^{1-\lambda}-1]}\mathrm{d}s]^{-1}.$$

In particular, if $0\leqslant\lambda<1$, $z_0z_1<0$ and $1+z_0^{-1}z_1\int_0^{+\infty}e^{\frac{\mu}{\lambda-1}[(1+s)^{1-\lambda}-1]}\mathrm{d}s<0$ or $\lambda>1$, $z_0z_1<0$, then there exists t_0 which satisfies $t_0=z_0+z_1\int_0^{t_0}e^{\frac{\mu}{\lambda-1}[(1+s)^{1-\lambda}-1]}\mathrm{d}s=0$ such that $\lim\limits_{t\to t_0^-}z(t)=0$. According to the existence theorem of ordinary differential equation, $d'(t)$ must exist on $t\in[0,t_0)$ and

$$d'(t)=c(t)^2e^{\frac{\mu}{\lambda-1}[(1+t)^{1-\lambda}-1]}\left[d_1-\frac{z_1}{c(t)}+a\int_0^t e^{\frac{\mu}{1-\lambda}[(1+s)^{1-\lambda}-1]}\mathrm{d}s\right],$$

$$d(t)=d_0+\int_0^t c^2(s)e^{\frac{\mu}{\lambda-1}[(1+s)^{1-\lambda}-1]}\left[d_1-\frac{z_1}{c(s)}+\int_0^s ae^{\frac{\mu}{1-\lambda}[(1+\xi)^{1-\lambda}-1]}\mathrm{d}\xi\right]\mathrm{d}s,$$

where, k,d_0,d_1 are constants. It is not difficult to find that there is $x_0\in\mathbb{R}$, such that

$$\lim_{t\to t_0^-}\rho(x_0,t)=+\infty.$$

Otherwise, the above $(c(t),d(t))$ is the global solution of the system of differential equations (3.1), so the analytical solution (2.1) is the global solution of the system (1.1).

4. Applications to other models

When $a=0$, systems (1.1) are the pressureless Euler equations with time-dependent damping. In this section, Theorem 2.1 is applied to the following initial boundary value problem

Theorem 4.1 For the pressureless Euler equations

$$\begin{cases} \rho_t+(\rho u)_x=0,(x,t)\in\mathbb{R}^+\times\mathbb{R}^+, \\ (\rho u)_t+(\rho u^2)_x=-\dfrac{\mu}{(1+t)^\lambda}\rho u,(x,t)\in\mathbb{R}^+\times\mathbb{R}^+, \\ \rho(x,0)=e^{-x},u(x,0)=x. \end{cases} \tag{4.1}$$

Where $\mu>0$. There exists an exact solution (ρ^*,u^*) with the form

(i) if $\mu=1,\lambda=1$,

$$\rho^*(x,t)=\frac{e^{-[1+\log(1+t)]^{-1}x}}{1+\log(1+t)},u^*(x,t)=\frac{x}{[1+\log(1+t)](1+t)}.$$

Moreover, for any fixed $x\in\mathbb{R}^+$, we have

$$\lim_{t\to+\infty}\rho^*(x,t)=0,\ \lim_{t\to+\infty}u^*(x,t)=0;$$

(ii)if $\mu\neq1,\lambda=1$,

$$\rho^*(x,t)=\frac{(1-\mu)e^{(\mu-1)[(1+t)^{1-\mu}-\mu]^{-1}x}}{(1+t)^{1-\mu}-\mu},u^*(x,t)=\frac{(1-\mu)x}{[(1+t)^{1-\mu}-\mu](1+t)^\mu}.$$

Moreover, for any fixed $x\in\mathbb{R}^+$, we have

$$\lim_{t\to+\infty}\rho^*(x,t)=\begin{cases}0,\mu<1\\ \dfrac{\mu-1}{\mu}e^{\frac{(1-\mu)x}{\mu}},\mu>1\end{cases},$$

$$\lim_{t\to+\infty}u^*(x,t)=0;$$

(iii)if $\lambda\neq1$,

$$\rho^*(x,t)=c(t)e^{-c(t)x},u^*(x,t)=c(t)xe^{\frac{\mu}{\lambda-1}[(1+t)^{1-\lambda}-1]},$$

where, $c(t)=\left(1+\int_0^t e^{\frac{\mu}{\lambda-1}[(1+s)^{1-\lambda}-1]}\mathrm{d}s\right)^{-1}$. Moreover, for any fixed $x\in\mathbb{R}$, we have

$$\lim_{t\to+\infty}\rho^*(x,t)=\begin{cases}Ae^{-Ax},\lambda<1\\ 0,\lambda>1\end{cases},$$

$$\lim_{t\to+\infty}u^*(x,t)=0;$$

where, $A=\left(1+\int_0^{+\infty} e^{\frac{\mu}{\lambda-1}[(1+s)^{1-\lambda}-1]}\mathrm{d}s\right)^{-1}$.

Journal of Physics: Conference Series **2543** (2023) 012007 doi:10.1088/1742-6596/2543/1/012007

References

[1] Courant R, Friedrichs K O. Supersonic flow and shock waves~[M]. Springer Science, 1999.
[2] Li D, Yu W. Boundary value problems for quasilinear hyperbolic systems~[M]. Duke University, 1985.
[3] Dafermos C M, Dafermos C M. Hyperbolic conservation laws in continuum physics~[M]. Berlin: Springer, 2005.
[4] Chen G, Pan R, Zhu S. Singularity formation for the compressible Euler equations~[J]. SIAM Journal on Mathematical Analysis, 2017,49(4):2591-2614.
[5] Hsiao L, Liu T P. Convergence to nonlinear diffusion waves for solutions of a system of hyperbolic conservation laws with damping~[J]. Communications in mathematical physics, 1992,143:599-605.
[6] Geng S, Wang Z. Convergence rates to nonlinear diffusion waves for solutions to the system of compressible adiabatic flow through porous media~[J]. Communications in Partial Differential Equations, 2010,36(5):850-872.
[7] Liu T P, Yang T. Compressible Euler equations with vacuum~[J]. Journal of differential equations, 1997,140(2):223-237.
[8] Pan X. Global existence of solutions to 1-d Euler equations with time-dependent damping~[J]. Nonlinear Analysis, 2016,132:327-336.
[9] Pan X. Remarks on 1-D Euler equations with time-decayed damping~[J]. arXiv preprint arXiv:1510.08115,2015.
[10] Pan X. Blow up of solutions to 1-d Euler equations with time-dependent damping~[J]. Journal of Mathematical Analysis and Applications, 2016,442(2):435-445.
[11] Sugiyama Y. Singularity formation for the 1D compressible Euler equations with variable damping coefficient~[J]. Nonlinear Analysis, 2018,170:70-87.
[12] Sugiyama Y. Remark on global existence of solutions to the 1D compressible Euler equation with time-dependent damping~[J]. Adv. Stud. Pure Math, 2020,85:379.
[13] Dong J, Xue H, Lou G. Singularities of solutions to compressible Euler equations with damping~[J]. European Journal of Mechanics-B/Fluids, 2019,76:272-275.
[14] Chen S, Li H, Li J, et al. Global and blow-up solutions for compressible Euler equations with time-dependent damping~[J]. Journal of Differential Equations, 2020,268(9):5035-5077.
[15] Dong J, Li J. Analytical solutions to the compressible Euler equations with time-dependent damping and free boundaries~[J]. Journal of Mathematical Physics, 2022,63(10):101502.

Journal of Physics: Conference Series **2543** (2023) 012008 doi:10.1088/1742-6596/2543/1/012008

Mathematical Modeling of pedestrian flow through the obstacle bottleneck

Xiaolu Jia[1,*], Daichi Yanagisawa[1,2,3], Claudio Feliciani[1], Katsuhiro Nishinari[1,2,3]

[1]Research Center for Advanced Science and Technology, The University of Tokyo, 4-6-1, Komaba, Meguro-ku, Tokyo 153-8904, Japan
[2]Department of Aeronautics and Astronautics, Graduate School of Engineering, The University of Tokyo, 7-3-1 Hongo, Bunkyo-ku, Tokyo 113-8656, Japan
[3]Mobility Innovation Collaborative Research Organization, The University of Tokyo, 5-1-5, Kashiwanoha, Kashiwa-shi, Chiba, 277-8574, Japan

* Corresponding author's e-mail: thexiaolujia@hotmail.com

Abstract. For pedestrians in walking facilities, their movements are often obstructed by bottlenecks where the walkable widths are geometrically reduced. In previous research, to reproduce the influence of bottlenecks on pedestrian movements, agent-based simulation models have been widely applied. However, their high reliability on modeling rules and parameters requires calibration at the microscopic level, which makes them difficult to apply from an engineering perspective. Here, we applied a mathematical approach, which estimates the egress efficiency based on the density-flow rate fundamental diagram, to reproduce the egress time of pedestrians in our field experiments. Both the obstacle and the exit were considered bottlenecks in our experiments. It was indicated that with the same width, the obstacle bottleneck acted as an 'ineffective' bottleneck that did not affect the egress time when it was near the exit bottleneck, whereas acted as an 'effective' bottleneck when it was distant from the exit bottleneck. To reproduce this phenomenon, we applied a mathematical approach that abstracts the walking scenario as a scheme with the bottlenecks as links and different regions as nodes. As a result, the egress times under different layouts were reproduced successfully by introducing the density-flow rate fundamental diagram into the scheme. Furthermore, a reasonable range of the obstacle size and obstacle-exit width, under which condition the egress time is constant, was estimated. This study can be applied to estimate the egress time of the walking facilities by considering the fundamentals of pedestrian flows from a mathematical perspective, thus helping in the actual design of bottlenecks that could ensure efficient and safe pedestrian egress in walking facilities.

1. Introduction

In walking facilities such as subway stations, sports venues, and commercial buildings, pedestrian movements are often affected by obstacles like walls, pillars, and interior furnishings. These obstacles often form bottlenecks that narrow pedestrian walkable space, thus affecting the physical [1, 2, 3, 4, 5, 6, 7, 8] and perceived congestion [9] of pedestrians. Therefore, exploring and reproducing the influence of the obstacle on pedestrian movement is significant to the actual geometrical design to guarantee a more comfortable, efficient, and safe walking environment.

In the case of pedestrian flow, agent-based models (a review is presented in [10]) have been widely used to emulate the influence of obstacle on pedestrian evacuation. Many studies have demonstrated the merits of obstacle before the exit in improving evacuation efficiency

[1, 2, 3, 4, 5]. One of the main reasons for the improvement of evacuation efficiency is that the obstacle could help decrease the conflicts caused by friction and turning behavior of pedestrians before the exit [1, 2, 11]. Other reasonable obstacle settings that could help improve evacuation efficiency have also been evaluated [12, 6]. On the other hand, some studies reminded us that under conditions where pedestrians cannot fully recognize the exit location, the obstacle could have a negative effect on evacuation efficiency by obstructing the sights of pedestrians [13]. Despite the fruitful results on agent-based modelings, the insufficient experimental evidence for modeling calibration makes some of these results lack of evidence.

As a result, many experimental studies have been conducted to provide evidence for modeling calibration. For instance, the obstacle–evading behavior of several pedestrians has been experimentally analyzed in [14, 15, 16] to provide evidence for the collision–avoidance rules of agent-based models. However, the calibration of agent-based modeling requires to be applied at a microscopic level should be time-consuming, let alone their sensitivity to parameters [17] that makes it challenging to apply these models into engineering practice.

To handle the artificiality of modeling, experiments have been conducted to provide evidence for the influence of obstacle bottleneck on pedestrian egress efficiency (a brief review seen in [8]). Under slow running and corner exit conditions, the obstacle was proved to be more helpful in increasing the evacuation efficiency than the normal walking and middle exit conditions [18]. It was also proposed that the efficiency would be improved more effectively if the obstacle was shifted from the exit center [1, 2]. The function of an obstacle to reducing conflicts has also been observed in other scenarios such as pedestrian intersections [19] where there were more conflicts among pedestrians. Other experiments showed that in crowded conditions with urgent participants, placing two obstacles before the exit would contribute to a higher evacuation efficiency compared with one obstacle or no-obstacle case [6]. On the other hand, some studies proposed that the obstacle only worked well on granular flow but barely worked for actual pedestrians, especially in highly crowded and competitive conditions [7]. Although the pedestrians were already very competitive in their experiments, the obstacle in their experiments could not reduce conflicts.

However, it is difficult to apply the experimental data to calibrate the parameters of agent-based modelings. For instance, the obstacle-evading behavior of individual pedestrians [15, 16] requires the calibration of seven parameters [14]. When it comes to the crowd, the calibration of agent-based modeling requires to be applied at the microscopic level which should be time-consuming or even unrealistic considering pedestrian heterogeneity. Let alone the sensitivity of these models to parameters [17] make it challenging to apply these models to engineering practice.

Here, to take advantage of experimental data to reproduce pedestrian flow, we apply a mathematical approach that estimates the pedestrian egress time based on the flow characteristics of pedestrians. We conducted field experiments in a room egress scenario with a wall-shaped obstacle, which formed another bottleneck aside from the exit bottleneck.

In the most present research, the obstacles were pillar-shaped and only placed before the exit. However, in actual situations, wall-shaped obstacles like partition walls and fences have also been widely applied, and the locations of obstacles are not limited to near the exit. Therefore, a more generalized study on the influence of a wall-shaped obstacle on pedestrian egress efficiency should be conducted. The pedestrian behavior at a single bottleneck has been widely explored at a normal exit bottleneck with a decrease in corridor width [20, 21, 22] or the conjunction of a T-shaped corridor [23, 24]. However, there are relatively few studies in the case of multiple bottlenecks. In a simulation study where bottlenecks have been added before the exit [25], it is shown that decreasing the flow at the extra bottlenecks would, in turn, improve the flow at the exit by decreasing the conflicts before the exit. Nevertheless, as to the authors' knowledge, there is no experimental study on the double-bottleneck situation caused by the obstacle.

The remainder of this paper is organized as follows. Section 2 gives details on field experiments with pedestrians, after which the flow characteristics are analyzed in Section 3. The calibration of modeling parameters is given in Section 4. Finally, a comparison of our mathematical approach and experiments is provided in Section 5, followed by the conclusions and discussions in Section 6.

2. Experiment setup

Experiments have been performed to explore the influence of obstacle size and location on the egress of crowded pedestrians. The experiments were conducted in December 2018 in the outdoor open space in front of the RCAST Building 4, The University of Tokyo, Japan. The experiment scenario was set as a corridor with a wall-shaped obstacle placed at the horizontal middle axis.

Totally 49 participants have joined our experiments. Among the participants, there were 29 males and 20 females whose ages ranged from 18 to 78 years old and whose heights ranged from 145 to 180 cm. We guaranteed the diversity of our participants to make our experiments more approximate to real egress cases with high heterogeneity of pedestrians.

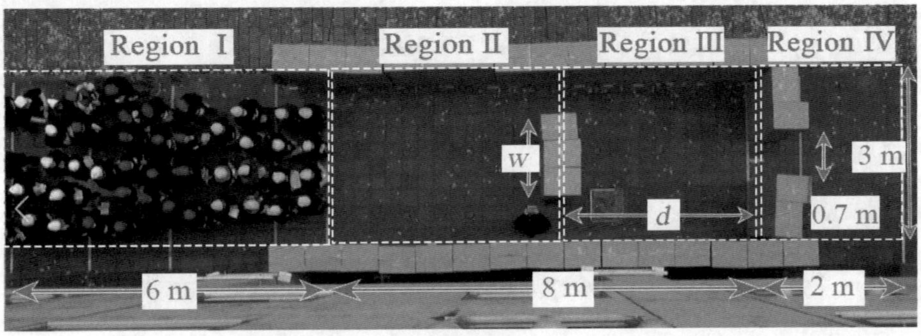

Figure 1. The geometrical layout of the experimental setting

As can be seen in Figure 1, the whole scenario is divided into Region I, II, III, and IV with the dashed lines as borders. Region I is the pedestrian waiting region that provided pedestrians with spaces to stand before each experiment test. Region II and III are respectively the walking regions before and after the obstacle. Pedestrians were required to walk straight in Region IV after passing by the exit in order not to impede the pedestrians in the corridor.

The obstacle width w and the obstacle-exit distance d were variable. Considering the obstacle was built with cardboard boxes, we would like to define the number of boxes that were used to build the obstacle as box. The width of each box is 0.42 m, and the obstacle width w can be calculated accordingly. The values of box, w, and d in each test can be seen in Table 1. All the experiments have been repeated at least twice.

Before each experiment test, participants were required to stand randomly in Region I. Afterward, they were instructed to start walking together, traverse Region II and III to egress from the exit, and keep walking straight in Region IV to avoid impeding other pedestrians. A camera was set above the horizontal axis of the corridor and fixed about 20 meters above the ground. Recordings of the camera were adjusted to 4k mode (3840×2160 pixel) with a frame rate of 30 fps. With the videos of the experiments as raw data, the recognition and tracking of pedestrians could be achieved using PeTrack software [26]. Pedestrians were required to wear colored caps and black shirts so that their positions in each video frame could be detected.

3. Influence of obstacle on egress efficiency in experiments

In this section, we would like to examine the influence of the obstacle on the egress efficiency by calculating the egress time. For a certain experiment run, we define t_i^{in} and t_i^{out} respectively

Table 1. The geometrical conditions of each test

d (m)	box	w (m)	repetitions
-	0	0	2
4	1	0.42	2
	2	0.84	2
	3	1.26	2
	4	1.68	2
1	1	0.42	2
	2	0.84	3
	3	1.26	2

as the moment for pedestrian i to get into Region II and leave Region III through the exit. Numbering the 49 pedestrians by the sequence to pass through the exit, we define the egress time T of a certain experiment run as the time for the first 46 pedestrians to traverse Region II and Region III. In other words, the egress time T is equal to the time lag between the first pedestrian stepping into Region II and the 46th pedestrian leaving Region III. Calculation of T can be seen in Equation 1:

$$T = t_j^{\text{out}} - t_i^{\text{in}}, \tag{1}$$

where $i = 1, j = 46$. We do not consider the last three pedestrians because there were no other pedestrians behind to affect them and may largely affect the total egress time. The variation of egress time T with obstacle width w under two types of obstacle-exit distance d can be seen in Figure 2. It can be seen from Figure 2 that when $d = 4$ m, the egress time T would roughly rise with the increase of obstacle size. We hence assume that the existence of the obstacle would decrease the egress efficiency when $d = 4$ m. By contrast, when $d = 1$ m, the egress time tends to keep constant despite the increase in obstacle size. Therefore, we assume that the existence of the obstacle does not affect the egress efficiency when $d = 1$ m. We could hence conclude that the influence of obstacle width on egress time is affected by obstacle-exit distance.

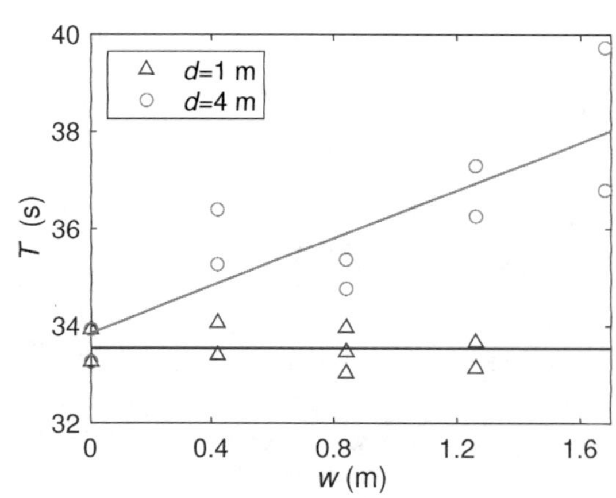

Figure 2. Variation of egress time under different obstacle sizes and locations

To explore the influencing mechanism of obstacle-exit distance on pedestrian egress, we would like to abstract the walking scenario in Figure 1 as a scheme. To be specific, the four regions could be considered nodes, and the bottlenecks are considered links between nodes. The pedestrian number at Region I, Region II, Region III, and Region IV are respectively defined as N_1, N_2, N_3 and N_4, and the flow rate, i.e., the number of pedestrians that pass by a section within one second, among different regions are respectively defined as Q_{in}, Q_{obs} and Q_{out}. The scheme is illustrated below:

$$\textcircled{N_1} \xrightarrow{Q_{\text{in}}} \textcircled{N_2} \xrightarrow{Q_{\text{obs}}} \textcircled{N_3} \xrightarrow{Q_{\text{out}}} \textcircled{N_4} \tag{2}$$

Together with the bottleneck at the exit, the existence of the obstacle makes the walking scenario a double-bottleneck environment. Initially, all 49 pedestrians were allocated in Region I. Afterward, pedestrians would move from Region I, pass by Region II and Region III, and finally leave the experimental region after passing by Region IV. With the variation of obstacle size w and obstacle-exit distance d, the Q_{obs} and Q_{out} would also variate, which might be the main reason for the variation of egress time under different conditions. The interpretation of Q_{obs} and Q_{out} under different w and d in our experiments can be seen in Figure 3.

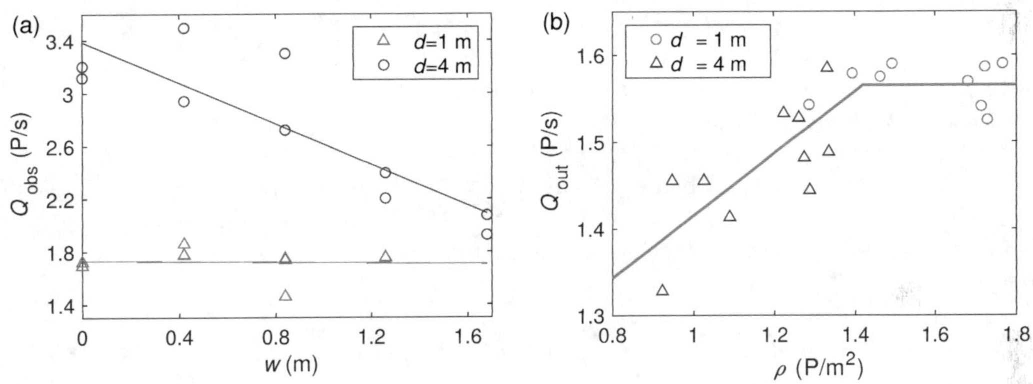

Figure 3. Variation of flow rate at the bottlenecks of (a) obstacle and (b) exit.

It can be seen in Figure 3(a) that the variation trend of Q_{obs} against obstacle width w would change under different obstacle-exit distance d. The dots represent the experimental data and the straight lines are the fitting curves to show the variation trend. It is shown that the Q_{obs} will roughly decrease with the rise of w under $d = 4$ m, and we define its corresponding value as $Q_{\text{obs}}^{\max}$. In contrast, under $d = 1$ m, the Q_{obs} tends to keep constant at a certain value despite the increase of obstacle width w. The equations of the fitting curves can be seen in Equation 3:

$$Q_{\text{obs}} = \begin{cases} Q_{\text{obs}}^{\max} = Aw + B, & \text{for} \quad d = 4\text{m}, \\ 1.73 \text{ P/s}, & \text{for} \quad \text{d} = 1\text{m}, \end{cases} \tag{3}$$

where $A = -0.77$ P/ms, $B = 3.39$ P/s. The different variation trends of Q_{obs} could indicate different effects of the obstacle on egress efficiency. As can be seen in Figure 3(a), the Q_{obs} is affected by obstacle size when $d = 4$ m, which means the obstacle forms an 'effective' bottleneck that could obstruct pedestrian flow. By contrast, when $d = 1$ m, the obstacle does not influence on the egress efficiency, which means the obstacle forms an 'ineffective' bottleneck that would not obstruct pedestrian egress.

We presume that the variation of flow rate at the exit could help explain our assumption about the 'ineffective' and 'effective' bottleneck. In Figure 3(b), ρ represents the average pedestrian density at Region III during one whole egress process. The dots are the experimental data showing the relation between ρ and Q_{out}. Through observation, we presume that Q_{out} tends to increase with ρ when $d = 4$ m, while tends to keep constant when $d = 1$ m. Accordingly, a piecewise linear function has been used to fit the dots and illustrated by straight lines in Figure 3(b). The variation of Q_{out} with ρ in the fitting function can be seen in Equation 4:

$$Q_{\text{out}} = \begin{cases} C\rho + D, & \text{for} \quad \rho < \rho_{\text{cri}}, \\ Q_{\text{out}}^{\text{max}}, & \text{for} \quad \rho \geq \rho_{\text{cri}}, \end{cases} \tag{4}$$

where $C = 0.35$ m^2/s, $D = 1.10$ P/s, $Q_{\text{out}}^{\text{max}} = 1.58$ P/s, $\rho_{\text{cri}} = 1.4$ P/m^2. It is shown that with the increase of average density ρ, the flow rate at the exit Q_{out} would first gradually increase, and then keep constant after the density reaches $\rho_{\text{cri}} = 1.4$ P/m^2. We presume the different increasing trends of egress time under different obstacle-exit distances could be explained by the variation of density. Under the same inflow rate Q_{obs}, the area of Region III is 12 m^2 when $d = 4$ m, and 3 m^2 when $d = 1$ m. Therefore, it would take a longer time to reach the critical density ρ_{cri} under the condition when $d = 4$ m. In other words, the duration when $\rho = \rho_{\text{cri}}$ is longer when $d = 1$ m. Since the Q_{out} under $\rho < \rho_{\text{cri}}$ is smaller than that when $\rho = \rho_{\text{cri}}$, the average Q_{out} under $d = 4$ m is smaller than that under $d = 1$ m.

4. Modeling the influence of the obstacle on pedestrian egress

According to our experimental results, we have discovered that the egress time is affected by both the obstacle width and obstacle-exit width. Based on the analytical results, we would like to build a mathematical model that could reproduce the experimental egress time under different obstacle layouts. With the mathematical model, we could estimate the egress time under a variety of obstacle layouts that are not limited by the obstacle layouts in our experiments. Furthermore, the estimation results of the model are expected to help guide the actual design of the obstacle.

4.1. The necessity of building a more complex mathematical model

We would like to first give the most common idea that can be usually thought of when calculating the egress time in the scheme shown in Equation 2, thus indicating the necessity of building a more complex model to reproduce the influence of obstacle layouts.

Generally, the outflow time could be calculated by dividing the volume by the flow capacity of the scheme. One of the basic assumptions of the Minimum-Cost Flow Problem, which has been widely used for static traffic assignment [27], indicates that the scheme capacity is equal to the minimum capacity of all the arrows. In other words, the arrow with the minimum capacity is the only bottleneck that impedes the flow. In our experiments, the exit width is 1 m while the least bottleneck width at the obstacle is 1.32 m. Therefore, the exit can be considered the main bottleneck, and the scheme capacity is equal to the capacity of the exit.

In this sense, to estimate the whole egress time of our experimental results, we presume the egress time could be separated into two periods. The first period is from the beginning to the first pedestrian reaches the exit, during which period there will be no outflow at the exit. The second period is from the moment the first pedestrian leaves the corridor to when the 46th pedestrian leaves the corridor. During the second period, pedestrians would consecutively egress from the exit. With the method above, the egress time in our experiments could be estimated as below.

In the first period, the duration is the time it costs for the first pedestrian to reach the exit. We assume the free flow velocity as $v = 1.5$ m/s and the walking distance as 8 m, i.e. the corridor length. The detour distance caused by the obstacle would be ignored because the detour distance would be mostly 0.2 m, which is neglectable compared with the corridor length.

Meanwhile, the obstacle would not affect pedestrian velocity under free-flow conditions [16]. As a result, the duration of the first period can be estimated as $8/1.5 \approx 5.33$ s.

In the second period, the common idea for calculation is to divide the total pedestrian number by the flow capacity at the exit. As is shown in Figure 3(b), the flow capacity at the exit is $Q_{\text{out}}^{\text{max}}$ =1.58 P/s. As a result, the duration of the second period, i.e. the period from the first pedestrian arriving at the exit to the 46th pedestrian leaves the corridor, can be estimated as $45/1.58 = 28.48$ s. Accordingly, the total egress time can be estimated as the sum of the two periods. The estimated egress time is 33.81 s, which follows the experimental egress time with no obstacle as shown in Figure 2.

Despite this simple calculation that could reproduce the egress time when there is no obstacle, the influence of obstacle width on egress time would be neglected because the bottleneck at the obstacle is never the main bottleneck in our experimental scenario. As a result, this simple estimation method could not be used to estimate the influence of the obstacle.

We presume the main reason that causes the accordance of estimation results with the experimental results is the assumption in the Minimum-Cost Flow Problem that the scheme capacity is only decided by the links. However, in our experiments, the capacity at the bottleneck is also affected by the pedestrian density within the nodes. To be specific, as is shown in Figure 3(b), the flow rate at Q_{out} is also affected by the pedestrian density ρ within Region III. Therefore, introducing the $\rho - Q_{\text{out}}$ relation in Region III should be developed to reproduce the influence of obstacle layout on egress time.

4.2. Three assumptions for calculation

To reproduce the $\rho - Q_{\text{out}}$ relation in Region III, we consider pedestrian egress as a dynamic process rather than static. Main parameters such as Q_{obs}, Q_{out} and ρ would be time-dependent in our calculation, which means they would change with time. Based on the variation of Q_{obs} and Q_{out} with obstacle layout in Figure 3, we have listed three assumptions for the convenience of calculation.

Assumption 1: At a certain timing t, the pedestrian density $\rho(t)$ at Region III would be time–dependent that would be affected by the variation of $Q_{\text{obs}}(t)$ and $Q_{\text{out}}(t)$.

For better illustration, we assume the area of Region III as S, which is equal to the product of the obstacle-exit distance d and corridor width 3 m. Meanwhile, the density in Region III can be calculated as $\rho(t) = N_3(t)/S$. The relations among $N_3(t)$, $Q_{\text{obs}}(t)$ and $Q_{\text{out}}(t)$ can be seen in Equation 5:

$$N_3(t) = \int_0^t (Q_{\text{obs}}(x) - Q_{\text{out}}(x))\, dx, \tag{5}$$

where x indicates any time during 0 and t and is used for integral calculation.

Assumption 2: when $\rho(t) < \rho_{\text{cri}}$, $Q_{\text{obs}}(t)$ equal to $Q_{\text{out}}^{\text{max}}$ in Equation 3. When $\rho(t) \geq \rho_{\text{cri}}$, the $Q_{\text{out}}(t)$ would be equal to the $Q_{\text{out}}(t)$.

When $\rho(t) < \rho_{\text{cri}}$, the obstacle acts as the main bottleneck, making the obstacle width the contributing factor to the $Q_{\text{obs}}(t)$. In this condition, we assume the value of $Q_{\text{obs}}(t)$ is equal to $Q_{\text{obs}}^{\text{max}}$ in Equation 3, which means $Q_{\text{obs}}(t)$ is only affected by obstacle width w. When $\rho(t) \geq \rho_{\text{cri}}$, the capacity of Region III is reached, the exit becomes the main bottleneck, and the value of $Q_{\text{obs}}(t)$ the same with $Q_{\text{out}}(t)$.

Assumption 3: $Q_{\text{out}}(t)$ would increase with the rise of $\rho(t)$ when $\rho(t) < \rho_{\text{cri}}$, while keep constant when $\rho(t) \geq \rho_{\text{cri}}$.

We presume the relation between $Q_{\text{out}}(t)$ and $\rho(t)$ in our model is the same as the experimental results in Equation 4. Please note that Q_{out} and ρ in Equation 4 are respectively the average exit flow rate and density during the whole process. Nevertheless, we assume that the variation trend also fits when the $Q_{\text{out}}(t)$ and $\rho(t)$ are time–dependent. To be specific, when $\rho(t) < \rho_{\text{cri}}$, the critical density in Region III is not reached. As a result, Q_{out} would gradually increase with the rise of $\rho(t)$ until the ρ_{cri} is reached. When $\rho(t) \geq \rho_{\text{cri}}$, the critical density in Region III is reached. As a result, pedestrians would keep the maximum outflow $Q_{\text{out}}^{\text{max}}$ at the exit.

4.3. Calculation of Egress time

According to the different status of $Q_{\text{obs}}(t)$, $Q_{\text{out}}(t)$ and $\rho(t)$ in Region III, we have divided the whole egress process into four processes as shown in Table 2. The total egress time T is the accumulation of the four duration as shown in Equation 6:

$$T = T_1 + T_2 + T_3 + T_4. \tag{6}$$

Assuming the total number of pedestrians as $N_1(0) = 49$, the egress process is terminated when the 46th pedestrian pass through the exit, which is consistent with the calculation of egress time in experiments.

Table 2. Four processes of the egress process.

Duration	$Q_{\text{obs}}(t)$	Q_{out}	$\rho(t)$	Description
T_1	≥ 0	$= 0$	$< \rho_{\text{cri}}$	From the beginning to when the first pedestrian leaves.
T_2	> 0	> 0	$< \rho_{\text{cri}}$	From $Q_{\text{out}} > 0$ to when $\rho(t) = \rho_{\text{cri}}$ or to when $Q_{\text{obs}}(t) = 0$.
T_3	> 0	> 0	$= \rho_{\text{cri}}$	The duration when $\rho(t) = \rho_{\text{cri}}$ is kept.
T_4	$= 0$	≥ 0	$< \rho_{\text{cri}}$	From when $\rho(t) < \rho_{\text{cri}}$ and $Q_{\text{obs}}(t) = 0$ to when $t \geq t_{46}^{\text{out}}$.

The duration of the four processes in Table 2 could be calculated by the relations among $Q_{\text{obs}}(t)$, $\rho(t)$ and $Q_{\text{out}}(t)$ in our three assumptions. A detailed explanation of the four processes and the calculation of T_1, T_2, T_3, and T_4 are as follows. Please note that the t in the following calculation only counts from the beginning of the corresponding period, which means $t = 0$ at a certain period indicates the beginning of that period.

4.3.1. ***Calculation of*** T_1 In the first period, the duration is from the beginning to when the first pedestrian passes the exit. In this period, $Q_{\text{obs}}(t)$ becomes nonzero when the first pedestrian enters into Region III and $Q_{\text{out}}(t)$ is always zero. Considering that the first pedestrian is not obstructed by any other pedestrians and could walk at his desired speed, we define the duration T_1 as the time for the first pedestrian to egress from the corridor. Again, we presume the free flow velocity as $v = 1.5$ m/s and the walking distance as 8 m and ignore the detour distance caused by the obstacle. Therefore, T_1 can be calculated in Equation 7:

$$T_1 = \frac{8\text{m}}{1.5\text{m/s}} \approx 5.33 \text{ s}. \tag{7}$$

Besides, we define the duration between the first pedestrian arriving at Region III to the first pedestrian leaving Region III as Δt, which could be calculated as $\Delta t = d/v$. During the period of Δt, $Q_{\text{obs}}(t) > 0$ while $Q_{\text{out}}(t) = 0$.

4.3.2. ***Calculation of*** T_2 In the second period, both the inflow and outflow exist in Region III, i.e. $Q_{\text{obs}}(t) > 0$ and $Q_{\text{out}}(t) > 0$. Considering the bottleneck width at the obstacle is always larger than the exit width, the inflow $Q_{\text{obs}}(t)$ is always higher than the outflow $Q_{\text{out}}(t)$. As a result, $Q_{\text{out}}(t)$ and $\rho(t)$ would gradually increase. During this process, according to **Assumption 2**, the inflow should be constant as $Q_{\text{obs}}^{\max}$ that is not related to time t in this period. Nevertheless, the outflow $Q_{\text{out}}(t)$ should be time-dependent due to the variation of $\rho(t)$.

This period will terminate only if one of the two following conditions is satisfied.

Condition ①: when all the pedestrians, i.e. $N_1(0) = 49$ P, have passed by the obstacle. At the end of this period, the $Q_{\text{obs}}(T_2) = 0$ and $\rho(T_2) < \rho_{\text{cri}}$.

Condition ②: the critical density in Region III is reached. At the end of this period, the $\rho(T_2) = \rho_{\text{cri}}$ and $Q_{\text{obs}}(T_2) > 0$.

We would like to calculate T_2 by considering the two conditions. If this period is terminated by Condition ①, we assume the duration as T_2^1. If this period is terminated by Condition ②, we assume the duration as T_2^2. The relation among T_2 would equal to the smaller one between T_2^1 and T_2^2 as shown in Equation 8:

$$T_2 = \min(T_2^1, T_2^2). \tag{8}$$

In Condition ①, the duration is only decided by the total pedestrian number and the flow rate at the obstacle. Therefore, T_2^1 can be calculated by Equation 9:

$$T_2^1 = \frac{N_1(0)}{Q_{\text{obs}}^{\max}}. \tag{9}$$

In Condition ②, the duration is affected by time-dependent parameters including $N_3(t)$, $Q_{\text{out}}(t)$ and $\rho(t)$. According to the relations among the parameters in our three assumptions, define $N_3(t)$ in Equation 10:

$$N_3(t) = \int_0^{\Delta t + t} Q_{\text{obs}}(x)\,dx - \int_0^t Q_{\text{out}}(x)\,dx, \tag{10}$$

where $Q_{\text{obs}}(t) = Q_{\text{obs}}^{\max}$, and $Q_{\text{out}}(t)$ in Equation 11 as

$$Q_{\text{out}}(t) = C\rho(t) + D = \frac{CN_3(t)}{S} + D = Q_{\text{out}}^{\max}. \tag{11}$$

Equation 10 shows the pedestrian number at Region III, i.e. $N_3(t)$, which is the difference between the accumulated inflow and outflow. Equation 11 shows the relation between $N_3(t)$ and Q_{out} based on the $\rho - Q_{\text{out}}(t)$ relation shown in Equation 4. To solve Equation 10–11, we first represent $N_3(t)$ with $Q_{\text{out}}(t)$ according to Equation 11. Afterwards, the $N_3(t)$ in Equation 10 could be replaced by a formula of $Q_{\text{out}}(t)$, making Equation 10 becomes an implicit function equation of $Q_{\text{out}}(t)$ as shown in Equation 12:

$$\frac{S}{C}(Q_{\text{out}}(t) - D) = \int_0^{\Delta t + t} Q_{\text{obs}}(x)\,dx - \int_0^t Q_{\text{out}}(x)\,dx, \tag{12}$$

where $Q_{\text{obs}}(t) = Q_{\text{obs}}^{\max}$.

T_2^2 can be obtained by solving Equation 13:

$$T_2^2 = -\frac{S}{C}\ln\left(\frac{Q_{\text{out}}^{\max} - Q_{\text{obs}}^{\max}}{\left(\frac{C}{S}\Delta t - 1\right)Q_{\text{obs}}^{\max} + D}\right), \tag{13}$$

and the solution is shown in Equation 14:

$$\begin{aligned} T_2 &= \min(T_2^1, T_2^2) \\ &= \min\left(\frac{N_1(0)}{Q_{\text{obs}}^{\max}}, -\frac{S}{C}\ln\left(\frac{Q_{\text{out}}^{\max} - Q_{\text{obs}}^{\max}}{\left(\frac{C}{S}\Delta t - 1\right)Q_{\text{obs}}^{\max} + D}\right)\right). \end{aligned} \tag{14}$$

For further calculation of T_3 and T_4, we assume the number of pedestrians that have passed by the obstacle as N_{left}, and the number of pedestrians that remain in Region III at the end of this period as N_{remain}.

In the case of N_{left}, considering the inflow Q_{obs} into Region III is always constant in this period, N_{left} can be calculated by the Equation 15.

$$N_{\text{left}} = \int_0^{\Delta t+t} Q_{\text{obs}}(x)\,dx = Q_{\text{obs}}^{\max}(\Delta t + T_2). \tag{15}$$

In the case of N_{remain}, calculation of N_{remain} are different under Condition ① and Condition ②. Under Condition ①, the critical density in Region III is not reached. In this condition, $N_{\text{remain}} = N_3(T_2^1)$, which can be calculated through substituting T_2^1 into Equation 10. Under Condition ②, the critical density in Region III is reached, which means $N_{\text{remain}} = \rho_{\text{cri}} S$. Accordingly, the value of N_{remain} can be derived as shown in Equation 16:

$$N_{\text{remain}} = \begin{cases} Q_{\text{obs}}^{\max}\Delta t + (\Delta t + \frac{S}{C}(D - Q_{\text{obs}}^{\max}))(\exp(-\frac{C}{S}T_2^1) - 1), & \text{for} \quad \rho(T_2) < \rho_{\text{cri}}, \\ \rho_{\text{cri}} S, & \text{for} \quad \rho(T_2) = \rho_{\text{cri}}. \end{cases} \tag{16}$$

4.3.3. ***Calculation of*** T_3 If the critical density in Region III could not be reached in the second period, i.e. $\rho(T_2) < \rho_{\text{cri}}$, the third period will not exist according to its definition. In this condition, $T_3 = 0$. If $T_3 \neq 0$, it means that $\rho(T_2) = \rho_{\text{cri}}$ and $N_{\text{remain}} = \rho_{\text{cri}} S$ at the end of the second period. Meanwhile, the critical density in Region III will be reached, and the third period is defined as the duration that the critical density is always reached in Region III. In other words, the relation $\rho(t) = \rho_{\text{cri}}$ and $Q_{\text{out}}(t) = Q_{\text{obs}}(t) = Q_{\text{out}}^{\max}$ should always be satisfied. Besides, the $N_3(t)$ would always be equal to N_{left} in this period.

The third period will terminate only when all the pedestrians, i.e. $N_1(0) = 49$ P, have passed by the obstacle. With a constant outflow, T_3 can be calculated by using Q_{obs} to divide the number of pedestrians that will pass by the obstacle in this period. To all the 49 pedestrians, according to Equation 15, N_{left} pedestrians have passed by the obstacle within the first and second periods, and the rest pedestrians will pass by the obstacle within the third period. Therefore, T_3 could be calculated in Equation 17.

$$T_3 = \begin{cases} 0, & \text{for} \quad \rho(T_2) < \rho_{\text{cri}}, \\ \dfrac{N_1(0) - N_{\text{left}}}{Q_{\text{out}}^{\max}}, & \text{for} \quad \rho(T_2) = \rho_{\text{cri}}. \end{cases} \tag{17}$$

4.3.4. ***Calculation of*** T_4 During the fourth period, no pedestrian would step into Region III, i.e. $Q_{\text{obs}}(t) = 0$, and the initial number of pedestrians at the beginning of this period is N_{remain}. Besides, as has been mentioned before, we will stop recording the time when the 46th pedestrian pass the exit, which means the calculation will stop when three pedestrians remain in the corridor. For calculation, we assume the time for N_{remain} pedestrians to leave Region III as T_4^1 and the time for the last three pedestrians to leave as T_4^2. The duration T_4 could be calculated by subtracting T_4^2 from T_4^1. The value of T_4^1 can be obtained by solving Equation 18–19:

$$N_3(t) = \int_0^t Q_{\text{out}}(x)\,dx = N_{\text{remain}}, \tag{18}$$

$$Q_{\text{out}}(t) = C\rho(t) + D = \frac{CN_3(t)}{S} + D, \tag{19}$$

Through calculating the Equation 18–19 in a way similar to the calculation of T_2^2 in Equation 10–11, the value of T_4^1, T_4^2 and T_4 could be calculated as shown in Equation 20–22:

$$T_4^1 = -\frac{S}{c}\ln\left(\frac{Q_{\text{out}}^{\max}}{\frac{C}{S}N_{\text{remain}} + D}\right), \tag{20}$$

$$T_4^2 = -\frac{S}{c}\ln\left(\frac{Q_{\text{out}}^{\max}}{3\frac{C}{S} + D}\right), \tag{21}$$

$$\begin{aligned} T_4 &= T_4^1 - T_4^2 \\ &= -\frac{S}{c}\ln\left(\frac{3C + DS}{CN_{\text{remain}} + DS}\right). \end{aligned} \tag{22}$$

According to Equation 7, 14, 17, and 22, the values of T_1, T_2, T_3 and T_4 could be obtained, and the total egress time T could be calculated based on Equation 6. Except for the constant parameters including A, B, C, D, $Q_{\text{out}}^{\max}$ and ρ_{cri}, the only two variable parameters that would affect T are S and $Q_{\text{obs}}^{\max}$, which are respectively decided by d and w. Therefore, given a certain w and d, the total egress time could be estimated.

5. Analysis of modeling results

Based on our mathematical model, we could estimate the egress time T under different values of w and d to reproduce the influence of obstacle size on egress time. To validate the model, we would like to compare the results of experiments and modeling. The obstacle layouts in our experimental settings can be seen in Table 1.

The comparison results can be seen in Figure 4. The horizontal axis represents the egress time T in our experiments and the vertical axis represents the T estimated by our model. The circular and triangle points respectively represent the T when $d = 4$ m and $d = 1$ m. The closer these points are to the $y = x$ axis, the higher the modeling accuracy is. It can be seen in Figure 4 that most data points fall into the 95% CI (confidence interval). Therefore, we presume our model is capable to reproduce the experimental egress time T.

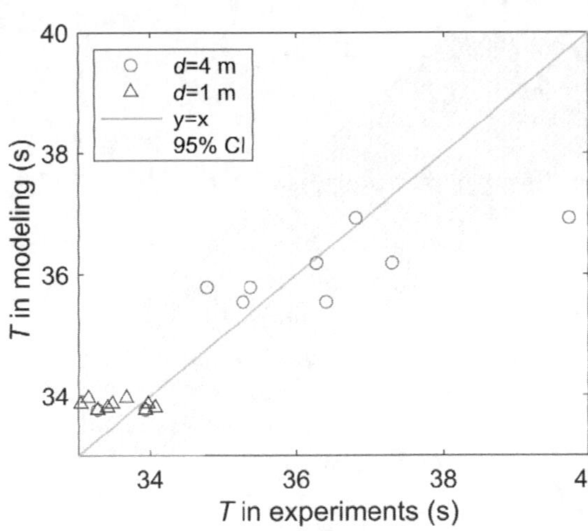

Figure 4. Comparison of egress time T obtained from experiments and modeling

After validating our model, we would like to calculate a more general relation between obstacle layouts, i.e. d and w, and egress time T. Compared with our experimental settings, we would like to extend the range of d as $1 \text{ m} \leq d \leq 6 \text{ m}$ and the range of w as $0.1 \text{ m} \leq w \leq 2.0 \text{ m}$. The calculation results of T under different w and d could be seen in Figure 5. The color bar is used to indicate the egress time T.

It can be seen in Figure 5 that T tends to increase with the rise of both w and d, which means both the increase of w and d would induce a longer egress time. To be specific, the increasing trend of egress time with obstacle size w is affected by the obstacle-exit distance d. When d is small, $d = 1$ m for instance, the increasing trend of T with w is relatively gentle. When d is large, $d = 6$ m for instance, the increasing trend of T with w is much more apparent. Therefore, it can be indicated that the larger the obstacle-exit distance is, the more apparent the increasing trend of egress time is with obstacle size.

We presume the influence of obstacle-exit distance is caused to the relation between pedestrian density and exit flow rate, i.e. $\rho - Q_{\text{out}}$ relation. Compared with a larger obstacle-exit distance, a smaller obstacle-exit distance would make the area before the exit filled quicker, thus causing a longer duration during which the exit flow rate keeps to the maximum.

Our presumption about the increasing trend of T is consistent with our experimental results. In our experiments, when $d = 1$ m, the increasing trend of T with w is not apparent, and the increasing trend could be fitted by a horizontal line. When $d = 4$ m, the increasing trend of T with w is more apparent and can be fitted by a linear function that rises with w.

Figure 5. Calculation results for the variation of egress time under different obstacle size w and obstacle-exit distance d

Additionally, we have depicted a critical curve which indicates the critical $d - w$ relation curve illustrated by the red critical curve in Figure 5. When the combination (w, d) is below the critical curve, the egress time is considered to be unapparent increased. According to our experimental results, the egress time is always the least when there is no obstacle with the average value being $T^{\text{min}} = 33.6$ s. We presume that the $T(w, d)$ is not different from T^{min} within a fluctuation of 5% as shown in Equation 23:

$$\frac{T(w,d) - T^{\text{min}}}{T^{\text{min}}} \leq 5\%, \tag{23}$$

which means that $T(w, d) \leq 35.28$ s in our experiments.

The critical curve could be fit by a quadratic polynomial function and the reasonable combinations of w and d can be seen in Equation 24:

$$d + 0.40 * w^2 + 0.25 * w \leq 4.04, \tag{24}$$

where $1 \text{ m} \leq d \leq 6 \text{ m}$, $0.1 \text{ m} \leq w \leq 2.0 \text{ m}$, and the corresponding $T(w, d) \leq 35.28$ s.

Results in Figure 5 and Equation 24 could help guide the setting of an obstacle in engineering applications. For instance, in the actual design of walking facilities, the obstacle would not increase the egress time only if the obstacle-exit distance and obstacle size could meet the reasonable range. Nevertheless, the calculation results are affected by the constant parameters including A, B, C, D, $Q_{\text{out}}^{\text{max}}$ and ρ_{cri}. In our model, all of the constant parameters are obtained from experimental data. It is a problem whether these constant parameters are still suitable when the geometrical size of the walking scenario is mainly different from our experiments.

6. Conclusion and Discussion

In this study, we have conducted both experimental and mathematical modeling methods to explore the influence of an obstacle on the egress time of pedestrians in normal conditions. By considering the density-flow fundamental diagram of actual pedestrian in the mathematical modeling, the egress times under different obstacle layouts in our experiments were reproduced successfully.

In our experiments, we built a corridor with a wall-shaped obstacle in the middle and changed the obstacle size and obstacle-exit distance to explore the corresponding influence. Results show that the influence of obstacle size on egress time is affected by the obstacle-exit distance. When the obstacle-exit distance is small, the obstacle size has no apparent influence on the egress time. However, when the obstacle-exit distance is large, the egress time would roughly increase with the obstacle size.

We presume the influence of obstacle-exit distance can be explained by abstracting the walking scenarios into a scheme and exploring the function of the obstacle as a bottleneck. When the obstacle is near the exit, the flow rate at the obstacle bottleneck is not affected by obstacle size. In this situation, the obstacle bottleneck is an 'ineffective' bottleneck that would guide pedestrians other than obstruct them. When the obstacle is far from the exit, the flow rate at the obstacle bottleneck would decrease with the increase in obstacle size. In this situation, the obstacle bottleneck is an 'effective' bottleneck that would reduce the egress time by impeding pedestrian movement.

It is interesting that in our experiments, although the flow rate at the exit is always the least, the exit can be an 'ineffective' bottleneck when the obstacle acted as an 'effective' bottleneck. However, in traditional calculation methods without considering the density-flow rate fundamental diagram, the 'effective' bottleneck should be the one with the least flow rate. In consequence, the traditional methods cannot reproduce the influence of obstacle layout on the egress time.

To further explore the influencing mechanism of obstacle layout behind the experimental data, we have built a mathematical model that can reproduce the influence of obstacle layout on egress time in our experiments. Moreover, the model can estimate the egress time under a wider range of obstacle width and obstacle-exit distance. As a result, reasonable combinations of obstacle size and obstacle-exit distance that would not increase the egress time have been obtained.

The advantage of the proposed mathematical method to estimate the egress time is clear. Compared with agent-based models whose results highly depend on a number of simulation parameters that are difficult to calibrate, this mathematical method is superior in its easy application only by measuring the flow fundamentals of pedestrians. Furthermore, we expect this mathematical method to be superior when the object region is large with a large number of pedestrians, which is time-consuming or even unavailable for agent-based modelings, whereas with little calculation burden for the mathematical method.

The results of this study are expected to help with the actual design of the obstacle in walking facilities. For instance, the obstacle in actual walking facilities should not affect the egress time. Our modeling results could help evaluate whether a certain obstacle layout is under the reasonable range and provide feasible schemes to improve the obstacle layout. Nevertheless, there are also some limitations in our mathematical model. The constant parameters for modeling might change under the geometrical settings of the walking scenario. For practical use, these constant parameters should be measured in the actual design of the obstacle in walking facilities. In our future work, we would like to explore the variation of the constant parameters with the change of geometrical sizes of the walking scenario, thus extending our mathematical model for a higher application value. In consequence, we expect to help verify and improve the obstacle design rules in the present design criterion of walking facilities with the extended model in the future.

Acknowledgements
This work was supported by the following Japanese grants: JSPS KAKENHI Grant No. JP21K14377, JP20K14992, JP21H01570 and JP21H01352, and JST-Mirai Program Grant No. JPMJMI17D4 and JPMJMI20D1, Japan. The experiments in this paper have been reviewed and approved by the Research Ethics Committee at The University of Tokyo, Japan (No. 18–200).

References

[1] Yanagisawa, D., Kimura, A., Tomoeda, A., Nishi, R., Suma, Y., Ohtsuka, K. & Nishinari, K. Introduction of frictional and turning function for pedestrian outflow with an obstacle. *Phys. Rev. E.* **80**, 036110 (2009,9)
[2] Yanagisawa, D., Nishi, R., Tomoeda, A., Ohtsuka, K., Kimura, A., Suma, Y. & Nishinari, K. Study on Efficiency of Evacuation with an Obstacle on Hexagonal Cell Space. *SICE Journal Of Control, Measurement, And System Integration.* **3**, 395-401 (2010)
[3] Zhao, Y., Li, M., Lu, X., Tian, L., Yu, Z., Huang, K., Wang, Y. & Li, T. Optimal layout design of obstacles for panic evacuation using differential evolution. *Physica A: Statistical Mechanics And Its Applications.* **465** pp. 175 - 194 (2017)
[4] Frank, G. & Dorso, C. Room evacuation in the presence of an obstacle. *Physica A: Statistical Mechanics And Its Applications.* **390**, 2135 - 2145 (2011)
[5] Shiwakoti, N., Shi, X. & Ye, Z. A review on the performance of an obstacle near an exit on pedestrian crowd evacuation. *Safety Science.* **113** pp. 54 - 67 (2019)
[6] Li, J., Jingyu, L., Chao, S., Sicong, Y. & Han, Z. Obstacle optimization for panic flow reducing the tangential momentum increases the escape speed. *Plos One.* **9** pp. 1 - 15 (2014)
[7] Garcimartín, A., Maza, D., Pastor, J., Parisi, D., Martín-Gómez, C. & Zuriguel, I. Redefining the role of obstacles in pedestrian evacuation. *New Journal Of Physics.* **20**, 123025 (2018,12)
[8] Jia, X., Murakami, H., Feliciani, C., Yanagisawa, D. & Nishinari, K. Pedestrian lane formation and its influence on egress efficiency in the presence of an obstacle. *Safety Science.* **144** pp. 105455 (2021)
[9] Jia, X., Feliciani, C., Murakami, H., Nagahama, A., Yanagisawa, D. & Nishinari, K. Revisiting the level-of-service framework for pedestrian comfortability: Velocity depicts more accurate perceived congestion than local density. *Transportation Research Part F: Traffic Psychology And Behaviour.* **87** pp. 403-425 (2022)
[10] Duives, D., Daamen, W. & Hoogendoorn, S. State-of-the-art crowd motion simulation models. *Transportation Research Part C: Emerging Technologies.* **37** pp. 193 - 209 (2013)
[11] Shiwakoti, N., Sarvi, M., Rose, G. & Burd, M. Consequence of Turning Movements in Pedestrian Crowds during Emergency Egress. *Transportation Research Record.* **2234**, 97-104 (2011)
[12] Karbovskii, V., Severiukhina, O., Derevitskii, I., Voloshin, D., Presbitero, A. & Lees, M. The impact of different obstacles on crowd dynamics. *Journal Of Computational Science.* (2018)
[13] Cristiani, E. & Peri, D. Robust design optimization for egressing pedestrians in unknown environments. *Applied Mathematical Modelling.* **72** pp. 553 - 568 (2019)
[14] Rudloff, C., Matyus, T., Seer, S. & Bauer, D. Can Walking Behavior be Predicted?: Analysis of Calibration and Fit of Pedestrian Models. *Transportation Research Record.* **2264**, 101-109 (2011)

[15] Parisi, D., Negri, P. & Bruno, L. Experimental characterization of collision avoidance in pedestrian dynamics. *Phys. Rev. E.* **94**, 022318 (2016,8)
[16] Jia, X., Feliciani, C., Yanagisawa, D. & Nishinari, K. Experimental study on the evading behavior of individual pedestrians when confronting with an obstacle in a corridor. *Physica A: Statistical Mechanics And Its Applications.* **531** pp. 121735 (2019)
[17] Moussaïd, M., Helbing, D. & Theraulaz, G. How simple rules determine pedestrian behavior and crowd disasters. *Proceedings Of The National Academy Of Sciences.* **108**, 6884-6888 (2011)
[18] Shi, X., Ye, Z., Shiwakoti, N., Tang, D. & Lin, J. Examining effect of architectural adjustment on pedestrian crowd flow at bottleneck. *Physica A: Statistical Mechanics And Its Applications.* **522** pp. 350 - 364 (2019)
[19] Lian, L., Mai, X., Song, W., Richard, Y., Wei, X. & Ma, J. An experimental study on four-directional intersecting pedestrian flows. *Journal Of Statistical Mechanics: Theory And Experiment.* **2015**, P08024 (2015)
[20] Hoogendoorn, S. & Daamen, W. Pedestrian Behavior at Bottlenecks. *Transportation Science.* **39**, 147-159 (2005)
[21] Daamen, W., Hoogendoorn, S. & Bovy, P. First-Order Pedestrian Traffic Flow Theory. *Transportation Research Record.* **1934**, 43-52 (2005),
[22] Hoogendoorn, S. & Daamen, W. Pedestrian Behavior at Bottlenecks. *Transportation Science.* **39**, 147-159 (2005)
[23] Shahhoseini, Z., Sarvi, M., Saberi, M. & Haghani, M. Pedestrian Crowd Dynamics Observed at Merging Sections: Impact of Designs on Movement Efficiency. *Transportation Research Record.* **2622**, 48-57 (2017),
[24] Qiu, G., Song, R., He, S. & Yin, W. The pedestrian flow characteristics of Y-shaped channel. *Physica A: Statistical Mechanics And Its Applications.* **508** pp. 199 - 212 (2018)
[25] Ezaki, T., Yanagisawa, D. & Nishinari, K. Pedestrian flow through multiple bottlenecks. *Phys. Rev. E.* **86**, 026118 (2012,8)
[26] Boltes, M. & Seyfried, A. Collecting pedestrian trajectories. *Neurocomputing.* **100** pp. 127 - 133 (2013)
[27] Ravindra, A., Thomas, M. & James, O. Network Flows: Theory, Algorithms, and Applications. (1988)

Using multi-flexible body dynamic analysis and the Taguchi method to decrease the variation of output angular velocity in a double circular-arc harmonic gear drive

J C Ruan[1], C K Lee[2,*], C Y Wang[1], Y C Tsai[1] and H H Sung[1]

[1] Department of Mechanical Engineering, Cheng Shiu University, 840 Chengcing Rd., Kaohsiung, Taiwan (R.O.C.)
[2] Department of Industrial Engineering and Management, Cheng Shiu University, 840 Chengcing Rd., Kaohsiung, Taiwan (R.O.C.)

* Corresponding author's e-mail: cklee@gcloud.csu.edu.tw

Abstract. The performance of a double circular-arc tooth harmonic gear drive, which offers advantages over traditional designs, such as lower stress during contact and increased fatigue life, was investigated in this research. The gear drive's meshing was simulated using dynamic analysis software RecurDyn, and a significant variation in output angular velocity was observed. To address this issue, experiments were planned using the Taguchi method with the $L_9(3^4)$ orthogonal table. The data was analyzed using techniques such as average value calculation, signal-to-noise ratio calculation, variance analysis, and response chart analysis to identify the control factors with the greatest impact on output angular velocity variation. It was found that an improved design with lower output angular velocity variation could be obtained by adjusting these control factors to optimal levels. The findings suggest that the transmission quality of double circular-arc harmonic gear drives can be enhanced by combining multi-flexible body dynamic analysis with the Taguchi method.

1. Introduction

Harmonic gear drives, also known as planetary harmonic drives or strain wave gears, are a type of gear drive that utilizes a unique waveform on the gear tooth profile to achieve high efficiency and accuracy. Harmonic gear drives have several advantages over traditional gear drives, including high efficiency, compact design, low noise and vibration, high torque capacity, high speed capability, and high reliability. Several researchers have proposed various design methods for harmonic gear drives. Xin [1] developed a design method for the double circular-arc tooth profile of the basic rack in harmonic drives, using research on the engagement principles of circular-arc tooth profiles in harmonic drives, for both flexspline and rigid spline. Chen et al. [2] examined the use of double circular-arc common-tangent tooth profile (DCTP) in harmonic drives (HD), and a complete method for designing the tooth profile and evaluating backlash in HD was presented to enable more teeth to participate in meshing. Yao et al. [3] proposed a novel complex cycloid tooth profile (TP) with a variable TP angle, which not only enhances the meshing state but also increases the fitting capacity of the cycloid. Shi et al. [4] designed and evaluated a novel type of harmonic gear drive, in which the flexspline is equipped with common tangent cycloid tooth profiles. Li et al. [5] proposed a design approach for a wave generator using a closed convex curve and an optimization model to minimize the flex spline's maximum circumferential stress. Results showed a significant reduction in stress compared to traditional elliptic wave generators. Wang et al. [6] proposed a new deformation model for a flexspline that accounts for forced deformation in the working area and free deformation in the non-working area, while maintaining a constant deformation shape during rotation. Tang et al. [7] developed a mathematical model for mixed lubrication in a harmonic drive with a double circular-arc tooth profile, considering factors such as normal load at the point of mesh, relative velocity, contact geometry, and surface roughness. Song et al. [8] proposed a double circular-arc tooth profile for harmonic gears using curve mapping and bidirectional conjugation. Tang et al. [9] proposed a tooth profile design method for harmonic drive based on conjugate equations and cam contour curve. The meshing performance was compared between non-conjugate and conjugate parameter driving design methods. Despite the many studies on harmonic gear drives, no research utilizing multi-flexible body dynamic analysis for harmonic gear drives has been reported. In this research, a double circular-arc rack with a helix angle of 0 degrees is used, and the tooth shapes of the flexspline and circular spline are derived using coordinate transformation and the theory of gearing [10]. Solidworks software is used to create a 3D physical model of the entire double circular-arc harmonic gear drive. To investigate the output angular velocity of the flexspline, Recurdyn, a software for multi-flexible body dynamic analysis, is utilized. The Taguchi method is applied to decrease the output angular velocity variation. The control factors are adjusted to the optimal level for each factor to obtain an improved design. Comparing with the original design, the improved design effectively increases the stability of the harmonic drive.

2. Using multi-flexible body dynamic analysis for a double circular-arc harmonic gear drive

In this study, we examine a double circular-arc harmonic gear drive consisting of a wave generator with two rollers, an outer-toothed flexspline, and an inner-toothed circular spline, as illustrated in Figure 1. Because the teeth on the flexspline and circular spline have double circular-arc profiles, the harmonic gear drive is named double circular-arc harmonic gear drive. The geometric solid models for the gear drive were created using SolidWorks. The circular spline is fixed while the wave generator is driven at a constant angular velocity and the flexspline is subjected to a constant torque. To examine the output angular velocity of the flexspline, multi-flexible body dynamic analysis software, RecurDyn, is used. Since the flexspline is a flexible component, it is divided into finite elements as depicted in Figure 2. The wave generator rollers and the circular spline are treated as rigid bodies.

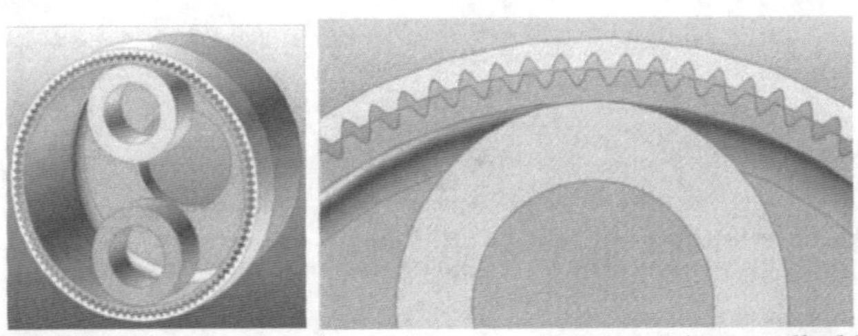

Figure 1. The double circular-arc harmonic gear drive studied herein.

Figure 2. The flexspline, which has been divided into finite elements.

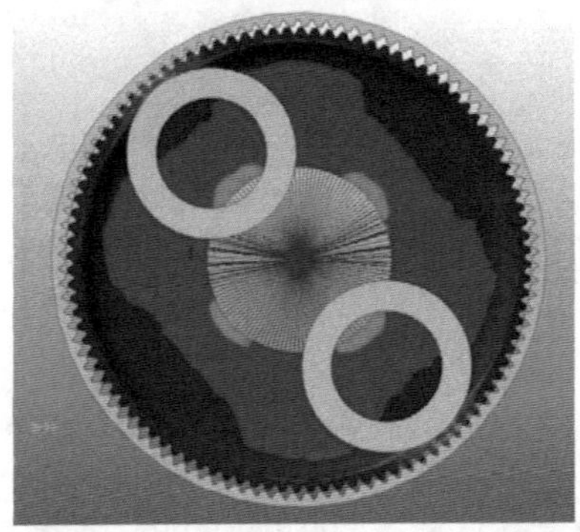

Figure 3. A moment in gear meshing.

After defining the material properties of the flexspline, establishing the contact conditions between surfaces, setting the connection joints, and applying the boundary conditions, the pre-processing can be completed. By using RecurDyn, the simulation of meshing of the double circular-arc harmonic gear drive is completed as shown in Figure 3. The output angular velocities of the flexspline over time for the load torques of 0 and 5000 N-mm are shown in Figures 4 and 5, respectively. When the load torque is 0, the difference between the maximum and minimum output angular velocity is 0.81083 rad/s. When the load torque is 5000 N-mm, the difference between the maximum and minimum output angular velocity is 0.48241 rad/s. To reduce the variation of output angular velocity, it is necessary to improve the design.

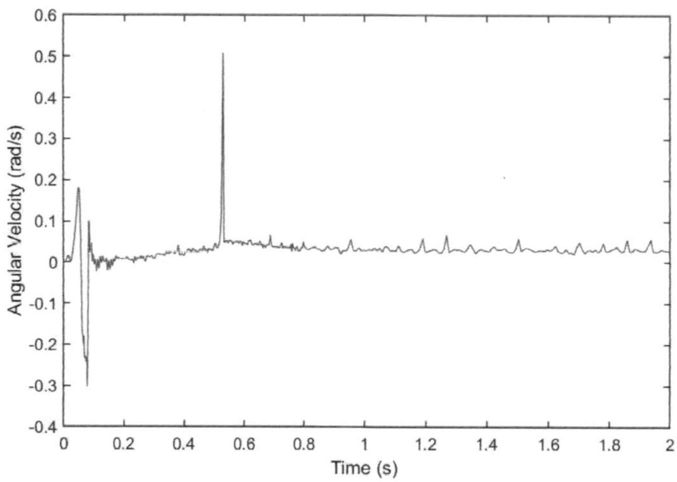

Figure 4. The graph of output angular velocity versus time for the original design when the load torque is 0 N-mm.

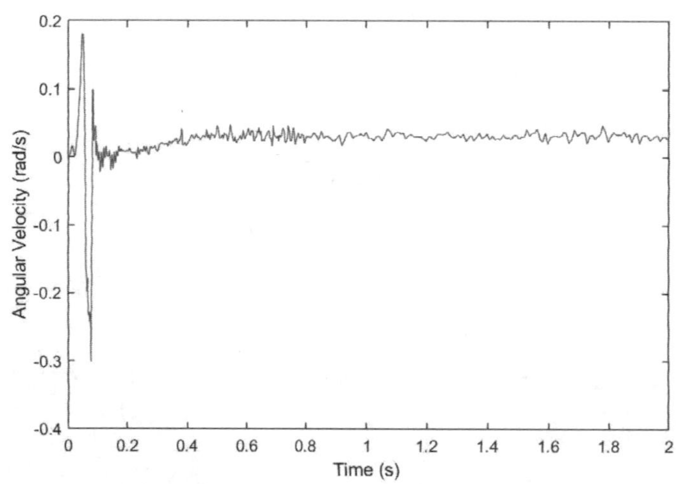

Figure 5. The graph of output angular velocity versus time for the original design when the load torque is 5000 N-mm.

3. Using the Taguchi method to improve the variation of output angular velocity

This paper uses the Taguchi method to minimize the variation of output angular velocity. Three dimensions are chosen as control factors, namely, the wall thickness of the flexspline (A), the height of the flexspline (B), and the diameter of the wave generator roller (C), as shown in Table 1. Each control factor has three levels. The $L_9(3^4)$ orthogonal array is used to generate nine experiments, as shown in Table 2, each representing a design configuration. Two load torque cases are considered, as shown in Table 3. Y1 represents the maximum variation of output angular velocity when the load torque is 0, Y2 represents the maximum variation of output angular velocity when the load torque is 5000 N-mm, AVG represents the average of Y1 and Y2, and SNR represents the smaller the better signal-to-noise ratio of Y1 and Y2. The ANOVA tables for SNR and AVG are shown in Tables 4 and 5. The response charts for SNR and AVG are shown in Figures 6 and 7. It can be seen that control factors A and C have a greater impact on SNR and AVG, while control factor B has a smaller impact on SNR and AVG. The optimal condition for SNR is A3B2C1, and the optimal condition for AVG is A3B1C1. Since factor B has a negligible impact on SNR and AVG, the level of factor B can be chosen arbitrarily. Therefore, the improved design is A3B2C1, which happens to be the 8th experiment. Figure 8 shows the output angular velocity over time for the improved design when the load torque is 0, and Figure 9 shows the output angular velocity over time for the improved design when the load torque is 5000 N-mm. Table 6 shows a comparison of the original design and the improved design. The improvement rate of AVG is 45.32% and the improvement rate of SNR is 156.88%. Increasing the thickness of the inner wall of the flexspline and reducing the diameter of the roller of the wave generator will help to improve the stability of the output angular velocity of the flexspline. Based on 3D printing technology, the improved design of double circular-arc harmonic gear drive has been made into a physical object. A video showcasing its operation can be found on the website https://youtu.be/LvKw9WUqsCs.

Table 1. Meaning and level values of control factors.

Control Factor	Meaning	1	2	3
A (mm)	The wall thickness of the flexspline	0.35	0.40	0.45
B (mm)	The height of the flexspline	35	40	45
C (mm)	The diameter of the wave generator roller	18.5	19.0	19.5

Table 2. Nine experiments planned by using $L_9(3^4)$ orthogonal array.

Exp. No.	A (mm)	B (mm)	C (mm)
1	1 (0.35)	1 (35)	1 (18.5)
2	1 (0.35)	2 (40)	2 (19.0)
3	1 (0.35)	3 (45)	3 (19.5)
4	2 (0.40)	1 (35)	2 (19.0)
5	2 (0.40)	2 (40)	3 (19.5)
6	2 (0.40)	3 (45)	1 (18.5)
7	3 (0.45)	1 (35)	3 (19.5)
8	3 (0.45)	2 (40)	1 (18.5)
9	3 (0.45)	3 (45)	2 (19.0)

Table 3. Results of experiments.

Exp. No.	Y1 (rad/s)	Y2 (rad/s)	AVG (rad/s)	SNR (dB)
1	2.72442	1.05768	1.89105	-6.3049
2	3.67623	2.90110	3.28867	-10.4003
3	5.55530	3.15242	4.35386	-13.0962
4	3.14603	1.80148	2.47376	-8.1766
5	5.47659	4.03039	4.75349	-13.6396
6	3.42793	3.53240	3.48017	-10.8330
7	3.68841	1.58146	2.63493	-9.0594
8	0.35670	0.35038	0.35354	9.0308
9	1.57381	1.82186	1.69783	-4.6210

Table 4. ANOVA table for SNR.

Source	SS	DOF	Var	F	Confidence	Rho
A	158.3	2	79.15	3.60	87.24%	42.28%
C	128.12	2	64.06	2.91	83.43%	34.22%
Error	87.97	4	21.99			23.5%

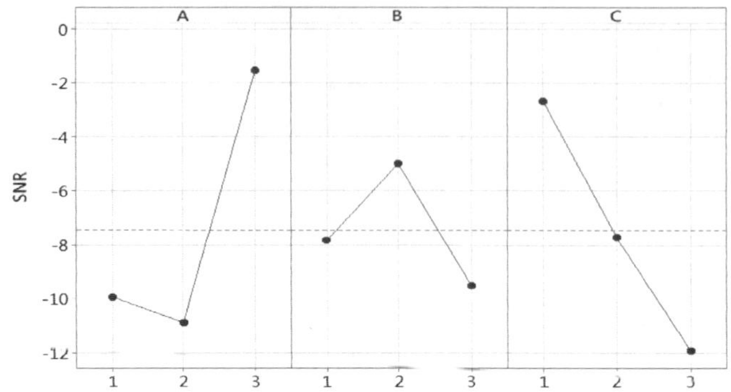

Figure 6. Response chart for SNR.

Table 5. ANOVA table for AVG.

Source	SS	DOF	Var	F	Confidence	Rho
A	6.79	2	3.40	7.16	95.24%	45.03%
C	6.40	2	3.20	6.75	94.77%	42.40%
Error	1.90	4	0.47			12.57%

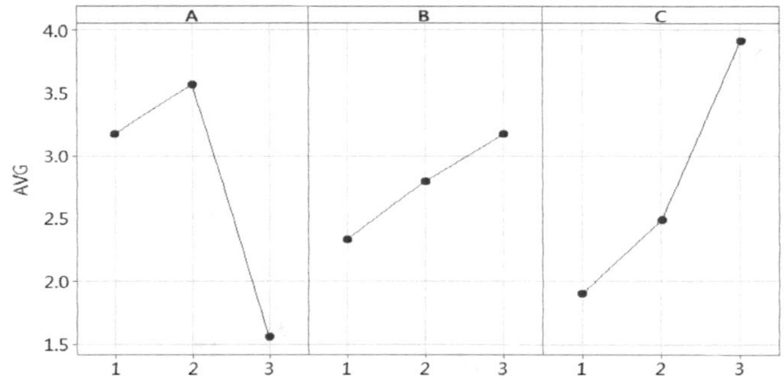

Figure 7. Response chart for AVG.

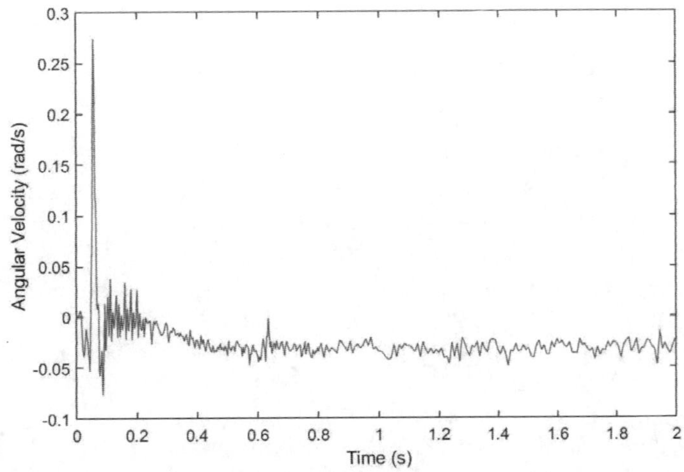

Figure 8. The graph of output angular velocity versus time for the improved design when the load torque is 0 N-mm.

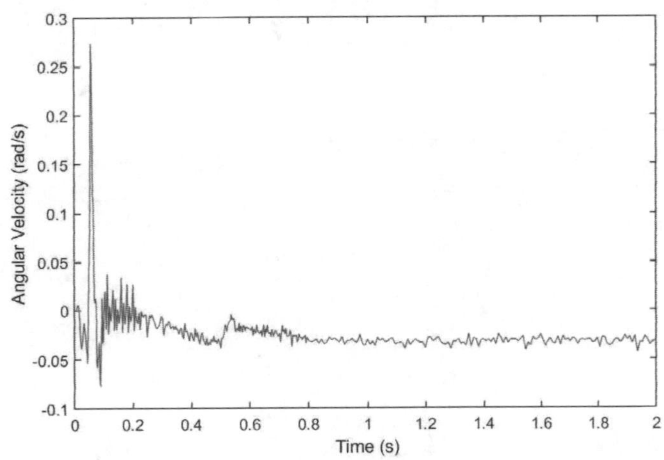

Figure 9. The graph of output angular velocity versus time for the improved design when the load torque is 5000 N-mm.

Table 6. Comparison of the original and improved designs.

Design	Condition	Y1 (rad/s)	Y2 (rad/s)	AVG (rad/s)	SNR (dB)
Original	A2B2C2	0.81083	0.48241	0.64662	3.5156
Improved	A3B1C1	0.35671	0.35038	0.35354	9.0308
Improvement Rate		56.01%	27.37%	45.32%	156.88%

4. Conclusion

This study aimed to reduce the variation of output angular velocity in a double circular-arc harmonic gear drive by using multi-flexible body dynamic analysis and the Taguchi experimental method. The results indicated that both the wall thickness of the flexspline and the diameter of the wave generator roller had a significant impact on the output angular velocity variation, while the height of the flexspline had a negligible effect. Increasing the wall thickness of the flexspline and decreasing the diameter of the wave generator roller improved the drive's stability. The average improvement rate was 45.32%, and the signal-to-noise ratio improvement rate was 156.88%. The drive demonstrated better stability under load torque than under no load. The study confirmed that the combination of multi-flexible body dynamic analysis and the Taguchi method can effectively enhance the transmission quality of the double circular-arc harmonic gear drive.

References

[1] Xin H 2011 *China Mech. Eng.* **22** 656-662
[2] Chen X, Liu Y, Xing J, Lin S and Xu W 2014 *Mech. Mach. Theory* **73** 1-24
[3] Yao Y P, Chen X X and Xing J Z 2017 *J. Ind. Prod. Eng.* **34** 1-8
[4] Shi W, Niu T and Song H 2018 *Mech. Eng. Technol.* **7** 43-53
[5] Li X, Song C, Yang Y, Zhu C and Liao D 2020 *Mech. Mach. Theory* **152** 103941
[6] Wang S, Li D, Mao S and Chen B 2021 *Appl. Sci.* **11** 6049
[7] Tang T, Li J, Wang J, Xiao K and Han Y 2022 *Proc. Inst. Mech. Eng. J.* **236** 480-498
[8] Song C, Li X, Yang Y and Sun J 2022 *Mech. Mach. Theory* **167** 104567
[9] Tang T, Li J, Wang J, Zeng X and Xiao K 2022 *J. Mech. Eng.* **58** 131-139
[10] Litvin F L and Fuentes A 2004 *Gear Geometry and Applied Theory (2nd ed.)* (Cambridge: Cambridge University Press)

Journal of Physics: Conference Series **2543** (2023) 012010 doi:10.1088/1742-6596/2543/1/012010

FDT for fluctuations of a special perturbated harmonic oscillator and its application in q-deformed Harmonic Oscillator

Huangcheng Yin

School of Mathematics and Physics, Xi'an Jiaotong-Liverpool University, Suzhou, China.

Corresponding author's e-mail: Huangcheng.Yin20@student.xjtlu.edu.cn

Abstract. In this paper, we will use fluctuation dissipation theorem on the harmonic oscillator with time correlation creation annihilation operator as the perturbated term and through calculation, we find that the exact value of the dissipation associated Imχ of this special perturbation harmonic oscillator is exactly equal to π. Next, we give its potential application in the q-deformed harmonic oscillator, which means connect the q-deformed harmonic oscillator with fluctuation dissipation.

1. Introduction

The fluctuation-dissipation theorem (FDT) is a well-established result, first formulated by Nyquist [1] and later proved by Callen and Welton [2]. It connects the fluctuations of the product of two operators with the dissipation expressed through the imaginary part of their response function. In this article, we will fluctuation dissipation of simple harmonic oscillator with time correlation creation annihilation operator as perturbation term, then calculate the susceptibility function and the average over fluctuation. During the calculation, it is found that Im($\chi(\omega)$) is a constant π. In the third part, we will discuss the practical application of this form of perturbation, which is an analogy of q-deformed Hamiltonian [3].

2. Fluctuation dissipation of simple harmonic oscillator with time correlation creation annihilation operator as perturbation term

The pure quantum harmonic oscillator can be described by the Hamiltonian $\hat{H}_0 = (\hat{a}^\dagger\hat{a} + 1/2)\hbar\omega$ where $\hat{a}^\dagger$, $\hat{a}$ correspond to the non-deformed annihilation and creation operators, respectively [4]. Let us consider the Hamiltonian perturbed by a term

$$\hat{H} = \hat{H}_0 + \hat{H}' = (\hat{a}^\dagger\hat{a} + 1/2)\hbar\omega - \hat{a}^\dagger$$

(Remark: It is not self-adjoint on account of $\hat{a}^\dagger(t)$. This term breaks the self-adjointness of the Hamiltonian and, therefore, nonphysical. However, we can see it's potential application in section 3.) $\hat{a}^\dagger(t)$ is an operator in the Heisenberg picture, related to its Schrodinger representation through

$$\hat{a}^\dagger(t) = exp(\frac{i\hat{H}_0 t}{\hbar})\hat{a}_s^\dagger exp(\frac{-i\hat{H}_0 t}{\hbar})$$

and $\hat{H}_0$ is the unperturbed Hamiltonian.

Under the condition $t \to -\infty \Rightarrow \hat{H}' \to 0$, expanding the Dyson series, and the eigenstates of the total Hamiltonian become

$$|\phi_n\rangle = |n\rangle - \frac{i}{\hbar}\int_{-\infty}^{t} dt'\, \hat{H}'(t')|n\rangle = |n\rangle - \frac{i}{\hbar}\int_{-\infty}^{t} dt'\, \hat{a}^{\dagger}(t')|n\rangle$$

where the last line corresponds to first-order perturbation theory, and $|n\rangle$ is a state of the unperturbed Hamiltonian with energy E_n (i.e., $\hat{H}_0|n\rangle = E_n|n\rangle$).
The expected value of another arbitrary operator $\hat{p}(t)$ is simply given by [5] [6]

$$\langle\hat{p}(t)\rangle = \frac{1}{Z}\sum_m \langle\phi_m|\hat{p}(t)|\phi_m\rangle\, e^{-\frac{E_m}{k_B T}}$$

$$\langle\hat{p}(t)\rangle = \frac{1}{Z}\sum_m \{\langle m|\hat{p}(t)|m\rangle + \frac{i}{\hbar}\int_{-\infty}^{t} dt'\, \varphi(t')\langle m|[\hat{p}(t),\hat{q}(t')]|m\rangle\} e^{-\frac{E_m}{k_B T}}$$

Where $Z = \sum_m e^{-\frac{E_m}{k_B T}}$ is the partition function at temperature T [7].
Thus in the above Hamiltonian, given the $\varphi = 1$, and has the intrinsic energy of the simple harmonic oscillator is $E_m = \hbar\omega(m + 1/2)$, and if let the another operator $\hat{p}(t) = \hat{a}(t)$, thus

$$\langle\hat{a}(t)\rangle = \frac{1}{Z}\sum_m \{\langle m|\hat{a}(t)|m\rangle + \frac{i}{\hbar}\int_{-\infty}^{t} dt'\, \langle m|[\hat{a}(t),\hat{a}^{\dagger}(t')]|m\rangle\} e^{-\frac{\hbar\omega(m+1/2)}{k_B T}}$$

The first term reduced to $\langle m|\hat{a}(t)|m\rangle = \langle m|\hat{a}_S|m\rangle$, recast it as

$$\langle\delta\hat{a}(t)\rangle = \langle\hat{a}(t) - \hat{a}(-\infty)\rangle = \int^{\infty} dt'\, \chi(t - t')$$

where susceptibility function is [5]

$$\chi(t - t') = \frac{i}{\hbar Z}\theta(t - t')\sum_m \langle m|[\hat{a}(t),\hat{a}^{\dagger}(t')]|m\rangle\, e^{-\frac{\hbar\omega(m+1/2)}{k_B T}}$$

Using the closure relation $|n\rangle\langle n| = I$, we can write

$$\langle m|[\hat{a}(t),\hat{a}^{\dagger}(t')]|m\rangle = \sum_n \{\langle m|\hat{a}_S|n\rangle\langle n|\hat{a}_S^{\dagger}|m\rangle - \langle m|\hat{a}_S^{\dagger}|n\rangle\langle n|\hat{a}_S|m\rangle\}\, e^{i\omega(m-n)(t-t')}$$

taking the time Fourier transform to work in frequency space, here we have

$$\int_0^{\infty} dt\, e^{i\omega' t} = \frac{i}{\omega' + i\epsilon}$$

Thus

$$\chi(\omega') = \int dt\, \chi(t) e^{i\omega' t}$$

After calculation, we can get

$$\chi(\omega') = \frac{-1}{Z}\sum_{m,n}\langle m|\hat{a}_S|n\rangle\langle n|\hat{a}_S^{\dagger}|m\rangle$$

In fact, by changing the frequency of the simple harmonic oscillator such that $\omega' = \omega$

$$\chi(\omega) = \frac{-1}{Z}\sum_{m,n}\langle m|\hat{a}_S|n\rangle\langle n|\hat{a}_S^{\dagger}|m\rangle\frac{e^{-\hbar\omega(m+1/2)/k_B T} - e^{-\hbar\omega(n+1/2)/k_B T}}{\hbar\omega(m - n + 1) + i\epsilon}$$

the dissipation associated to χ can be written

$$\mathrm{Im}(\chi(\omega)) = (1 - e^{-\hbar\omega/k_BT})\frac{\pi}{Z}\sum_{m,n}\langle m|\hat{a}_S|n\rangle\langle n|\hat{a}_S^\dagger|m\rangle\, e^{-\hbar\omega(m+1/2)/k_BT}\,\delta(m-n+1)$$

We can make reasonable settings $\hat{a}_S = \hat{a}$ and $\hat{a}_S^\dagger = \hat{a}^\dagger$ when $t = 0$ Thus

$$\mathrm{Im}(\chi(\omega)) = (1 - e^{-\hbar\omega/k_BT})\frac{\pi}{Z}\sum_{m}\langle m|\hat{a}|m+1\rangle\langle m+1|\hat{a}^\dagger|m\rangle\, e^{-\hbar\omega(m+1/2)/k_BT}$$

Notice the identity

$$\hat{a}|m+1\rangle = \sqrt{m+1}|m\rangle$$
$$\hat{a}^\dagger|m+1\rangle = \sqrt{m+1}|m+1\rangle$$

Insert the two identity in $\mathrm{Im}(\chi(\omega))$

$$\mathrm{Im}(\chi(\omega)) = (1 - e^{-\hbar\omega/k_BT})\frac{\pi}{Z}\sum_{m}(m+1)\, e^{-\hbar\omega(m+1/2)/k_BT}$$

Let's do some math skills, Multiply $e^{-\hbar\omega/2k_BT}$ on both numerator and denominator

$$\mathrm{Im}(\chi(\omega)) = (1 - e^{-\hbar\omega/k_BT})\frac{\pi}{e^{-\hbar\omega/2k_BT}Z}\sum_{m}(m+1)\, e^{-\hbar\omega(m+1)/k_BT}$$

Absorbing $m + 1$ into the differential form

$$\mathrm{Im}(\chi(\omega)) = (1 - e^{-\hbar\omega/k_BT})\frac{\pi}{e^{-\hbar\omega/2k_BT}Z}\sum_{m}\frac{k_BT}{-\hbar}\frac{de^{-\hbar\omega(m+1)/k_BT}}{d\omega}$$

$$\mathrm{Im}(\chi(\omega)) = (1 - e^{-\hbar\omega/k_BT})\frac{\pi}{e^{-\hbar\omega/2k_BT}Z}\frac{d\sum_m e^{-\hbar\omega(m+1)/k_BT}}{d\omega}\frac{k_BT}{-\hbar}$$

Let's look at the denominator now

$$e^{-\hbar\omega/2k_BT}Z = \sum_{\mathrm{m}} e^{-\hbar\omega(m+1)/k_BT}$$

Reverse Taylor expansion

$$\sum_{\mathrm{m}} e^{-\hbar\omega(m+1)/k_BT} = \mathrm{e}^{-\hbar\omega/k_BT}\frac{1}{1-\mathrm{e}^{-\hbar\omega/k_BT}} = \frac{1}{\mathrm{e}^{\hbar\omega/k_BT}-1}$$

Notice this is the Bose-Einstein distribution function [7]. Let's denote it as

$$\mathrm{n}(\omega) = \frac{1}{\mathrm{e}^{\hbar\omega/k_BT}-1}$$

Thus $\mathrm{Im}(\chi(\omega))$ can be written a

$$\mathrm{Im}(\chi(\omega)) = \frac{\mathrm{e}^{\hbar\omega/k_BT}-1}{\mathrm{e}^{\hbar\omega/k_BT}}\frac{\pi}{\mathrm{n}(\omega)}\frac{\mathrm{dn}(\omega)}{\mathrm{d}\omega}\frac{k_BT}{-\hbar} = \frac{1}{\mathrm{e}^{\hbar\omega/k_BT}}\frac{\pi}{n(\omega)^2}\frac{dn(\omega)}{d\omega}\frac{k_BT}{-\hbar}$$

$$\Rightarrow \mathrm{Im}(\chi(\omega)) = \frac{\pi}{\mathrm{de}^{\hbar\omega/k_BT}/\mathrm{d}\omega}\frac{\mathrm{d}\frac{1}{\mathrm{n}(\omega)}}{\mathrm{d}\omega}$$

$$\Rightarrow \mathrm{Im}(\chi(\omega)) = \pi\frac{\mathrm{d}\frac{1}{\mathrm{n}(\omega)}}{\mathrm{de}^{\hbar\omega/k_BT}} = \pi$$

Journal of Physics: Conference Series **2543** (2023) 012010 doi:10.1088/1742-6596/2543/1/012010

So under the such Hamiltonian, the $\mathrm{Im}(\chi(\omega))$ is a constant π. Notice the definition of the average over fluctuations is

$$S(t-t') = \langle \hat{a}(t)\hat{a}^\dagger(t')\rangle = \int \frac{d\omega}{2\pi} S(\omega) e^{-i\omega(t-t')}$$

$$S(\omega) = 2\hbar(n(\omega)+1)\mathrm{Im}(\chi(\omega))$$

Thus insert $\mathrm{Im}(\chi(\omega)) = \pi$ into $S(\omega)$

$$S(\omega) = 2\pi\hbar(n(\omega)+1) = h(n(\omega)+1)$$

where h is Planck constant.
Therefore, we can calculate the average value of the generated annihilation operators over time

$$\langle \hat{a}(\omega_1)\hat{a}^\dagger(\omega_2)\rangle = \int dt_1 dt_2\, e^{i\omega_1 t_1 + i\omega_2 t_2}\langle \hat{a}(t_1)\hat{a}^\dagger(t_2)\rangle = 2\pi\delta(\omega_1+\omega_2)S(\omega_1)$$

$$\langle \hat{a}(\omega_1)\hat{a}^\dagger(\omega_2)\rangle = 2\pi h(n(\omega_1)+1)\delta(\omega_1+\omega_2)$$

Using the same way, the Fourier transform of the fluctuation $\langle \hat{a}^\dagger(\omega_2)\hat{a}(\omega_1)\rangle$

$$\langle \hat{a}^\dagger(\omega_2)\hat{a}(\omega_1)\rangle = 2\pi h n(\omega_1)\delta(\omega_1+\omega_2)$$

Note that the amount we can observe is only Hermitian, and the symmetrized product is Hermitian.

$$\frac{1}{2}\langle \hat{a}(\omega_1)\hat{a}^\dagger(\omega_2) + \hat{a}^\dagger(\omega_2)\hat{a}(\omega_1)\rangle = 2\pi h(n(\omega_1)+\frac{1}{2})\delta(\omega_1+\omega_2)$$

Similarly, annihilation operator as perturbation term can be made

$$\hat{H} = \hat{H}_0 + \hat{H}' = (\hat{a}^\dagger\hat{a} + 1/2)\hbar\omega - \hat{a}(t)$$

3. Application of Hamiltonian of the above form

When a q-deformation is applied to the above system, the Hamiltonian is modified to the form [3]:

$$\hat{H}_q = \frac{1}{2}(\hat{A}^\dagger\hat{A} + \hat{A}\hat{A}^\dagger)$$

where $\hat{A}$ and its adjoint $\hat{A}^\dagger$ are the deformed annihilation and creation operators, respectively, the oscillator described by the Hamiltonian $\hat{H}_q$ is called the q-deformed harmonic oscillator.
The operators $\hat{A}$ and $\hat{A}^\dagger$ obey the deformed commutation relation

$$\hat{A}\hat{A}^\dagger - q^2\hat{A}^\dagger\hat{A} = I$$

where $0 < q < 1$.
I hope to use the fluctuation dissipation theorem, so rewrite the above form

$$\hat{H}_q = \frac{1}{1+q^2}\hat{H}_0 = \frac{1}{1+q^2}(\hbar\omega(\hat{a}^\dagger\hat{a} + \frac{1}{2}))$$

Perform Taylor expansion and remove the higher-order terms

$$\hat{H}_q = \frac{1}{1+q^2}\hat{H}_0 = \hat{H}_0 - q^2\hat{H}_0$$

Notice if q is much smaller than 1, $\hbar\omega q^2/2$ can be ignored. Thus

$$\hat{H}_q = \frac{1}{1+q^2}\hat{H}_0 = \hat{H}_0 - q^2\hbar\omega\hat{a}^\dagger\hat{a}$$

Notice $\hat{H}_q$ not equals $\hat{H}_q^\dagger$, thus this is also nonphysical. Therefore, we can consider the following disturbance forms:

$$\hat{H} = \hat{H}_0 - \lambda \hat{a}^\dagger(t)\hat{a}(t)$$

The operation steps are similar to deal with $\hat{H} = (\hat{a}^\dagger \hat{a} + 1/2)\hbar\omega - \hat{a}^\dagger$.

4. Conclusion

By focusing on $\hat{H} = (\hat{a}^\dagger \hat{a} + 1/2)\hbar\omega - \hat{a}^\dagger$, this paper finds that the results of this form of fluctuation dissipation are not complicated, and even unexpectedly simple and clear, at the same time, an application of this $\hat{H} = (\hat{a}^\dagger \hat{a} + 1/2)\hbar\omega - \hat{a}^\dagger$ is illustrated, that is, $-\lambda \hat{a}^\dagger(t)\hat{a}(t)$ simulated from the q-deformed Harmonic Oscillator.

5. Appendix

Kramers-Kronig Relations [5] [6]:

$$\chi_{B,A}(\omega) = \frac{-1}{Z}\sum_{m,n} \langle n|B|m\rangle\langle m|A|n\rangle \frac{e^{-E_n/k_BT} - e^{-E_m/k_BT}}{\hbar\omega + E_n - E_m + i\epsilon}$$

as $\chi_{B,A}(\omega)$ is also a complex-valued function along the real axis, so accordingly

$$\chi_{B,A}(\omega) = \chi'_{B,A}(\omega) + i\chi''_{B,A}(\omega)$$

And

$$\chi'_{B,A}(\omega) = \frac{1}{\hbar} PV \int d\omega' \frac{\chi''_{B,A}(\omega')}{\omega' - \omega}$$

$$\chi''_{B,A}(\omega) = -\frac{1}{\hbar} PV \int d\omega' \frac{\chi'_{B,A}(\omega')}{\omega' - \omega}$$

Here

$$PV\int d\omega' = \int_{-}^{\omega-0} d\omega' + \int_{\omega+0}^{\infty} d\omega'$$

is called the (Cauchy) principal value integral, and the above relations follow immediately from

$$\chi_{B,A}(\omega) = \frac{1}{\pi}\int d\omega' \frac{\chi''_{B,A}(\omega')}{\omega' - z} = \frac{1}{2\pi i}\int d\omega' \frac{\chi_{B,A}(\omega' + i\epsilon) - \chi_{B,A}(\omega' - i\epsilon)}{\omega' - z}$$

which in turn follows from the Residue Theorem because of the mentioned analytical and asymptotic properties of the susceptibility $\chi_{B,A}(\omega)$.

Reference

[1] Nyquist H 1928 *Phys. Rev.* **32**(110)
[2] Callen H B and Welton T A 1951 *Phys. Rev.* **83**(34)
[3] Eremin V V and Meldianov A A 2006 *Theor. Math. Phys.* **147**(709)
[4] Sakurai J J 2016 *Morden quantum mechanics* (Beijing: World Publishing Corporation)
[5] Czycholl G 2017 *Theoretische Festk orperphysik* Band 2 (Berlin: 4. Auflage, Springer Verlag)
[6] Czycholl G 2016 *Theoretische Festk orperphysik* Band 1 (Berlin: 4. Auflage, Springer Verlag)
[7] Feynman R P 1998 *Statistical Mechanics* (Westview: Advanced Book Classics)

Utilizing GOES-16 instruments to forecast geomagnetic storm during Solar Cycle 25

Anis Rossdee[1,2], Z S Hamidi[1,2,*] and N N M Shariff[2,3]

[1]Faculty of Applied Sciences, Universiti Teknologi MARA Shah Alam, 40450 Shah Alam, Selangor, Malaysia
[2]Institute of Science, Universiti Teknologi MARA Shah Alam, 40450 Shah Alam, Selangor, Malaysia
[3]Academy of Contemporary Islamic Studies (ACIS), Universiti Teknologi MARA Shah Alam, 40450 Shah Alam, Selangor, Malaysia

* Corresponding author's e-mail: zetysh@uitm.edu.my

Abstract. This paper focuses on the study of active region AR12891 in conjunction with geostorm from space weather data. A flare of magnitude M1.7 occurred in this active region, resulting in a geomagnetic storm of magnitude 3. By observing solar activity, x-ray flux, and radiation particle monitoring, GOES-16 instruments were used to forecast these events. A python program was used to analyze data collected from the online database of solar monitor and space weather live provided by the NOAA. Solar flare events can be studied and monitored through GOES-16 instruments to forecast geomagnetic storms. This allows less impacts towards satellite navigations and power sources if geomagnetic storm was detected early.

1. Introduction

Energetic particles emitted by the Sun could collide with the Earth's atmosphere Radio navigation, satellite communications, and the electrical grid may be affected by these effects. These impacts are not always caused by the sun, as they depend on the solar cycle (SC). As the Solar Cycle maximum approaches, there will be an increase in active regions (AR) that cause solar activity [1], flares will rise that release massive amounts of energetic particles that could interfere with Earth's magnetic field.

Radiation at frequencies ranging from radio waves to gamma rays could be emitted as a result of the energy released. Flares can be classified into A-, B-, C-, M-, and X- classes based on their intensity in X-rays (X-ray flux)[2]By releasing massive amounts of superheated gas into space, CMEs are vast portions of corona pushed into space by solar wind. In the event that coronal holes face Earth, their magnetic fields will affect the planet [3].

The GOES-16 weather satellite is operated by the National Oceanic and Atmospheric Administration (NOAA). Among these instruments are the Space Environment In-Situ Suite (SEISS), Extreme Ultraviolet and X-ray Irradiance Sensors (EXIS), and Magnetometer (MAG). Charged particles and radiation created by flares and CMES can collide with the magnetosphere in the solar wind, a plasma of ions and electrons. In the magnetosphere, SEISS monitors and detects proton, electron, and ion energy. In order to determine radiation risks to astronauts and spacecraft, experts use the SEISS to detect charged particles [4]

Using EXIS, solar irradiance is measured and tracked in the upper atmosphere. Earth's upper atmosphere can absorb a significant amount of electromagnetic energy that is emitted by the Sun [5]. Additionally, flares can interfere with satellite dragging, affecting communication operations, and blocking High Frequency (HF) radio communications. EXIS is capable of detecting X-ray from flares as it could interfere with navigational precision and lead to failure of power grid operation [6]

Meanwhile, MAG measures the magnetic field in the space environment[7], which is controlled by charged particles in the magnetosphere. The first sign that major space weather has reached or impacted the Earth's magnetic field is frequently found in magnetic observations[8]. These particles pose a threat to spacecraft. Measurements of the geomagnetic field can provide alerts and warnings to satellite operators. In this study, the GOES-16 instruments will be used to detect geomagnetic storms for the upcoming SC25 maximum, and to study the effects of geomagnetic storms on earth.

2. Methodology and data collection

Using GOES-16 instruments, data from the sun's solar activity was analyzed using coding sources [9] interpreted by Python programming language. The solar event selected was AR12891, which caused the most severe geomagnetic storm during SC25 in 2021. A live space weather prediction on the internet was used to get data on AR12891 and its associated flare. The GOES-16 Extreme Ultraviolet and X-ray Irradiance Sensors (EXIS) monitored the x-ray wavelengths released by a flare. To determine flare class intensity, plots of X-rays in the W/m^2 range were used. The SOHO (Solar and Hemispheric Observatory) spacecraft's LASCO C2 and C3 coronagraph sensors detected a CME using the helioviewer website. It is also possible to detect CME blasts using GOES-16 Space Environment In-Situ Suite (SEISS), a proton flux monitor that measures the quantity of 10 MeV protons near Earth's atmosphere. The GOES-16 magnetometer (MAG) was used to determine which particles would interact with the Earth's atmosphere following the incident. Geomagnetic activity statistics are provided by GOES-16 MAG. Radiation particles from solar flares can affect the magnetosphere and ionosphere, affecting communication and navigation.

3. Result and discussion

In this study, selected solar events were taken and observed to understand their impact on the earth's atmosphere. An M-class flare was released by AR12891 at 3:00UTC, followed by a full halo CME at 3:15UTC.

3.1. GOES-16 extreme ultraviolet and X-ray irradiance sensors (EXIS)

The X-ray sensor (XRS) on GOES-16 is split into two wavelength ranges, XRSa and XRSb, with wavelength ranges of 0.5-4.0 Armstrong and 1.0-8.0 Armstrong, respectively. Figure 1 illustrates a peak of x-ray flux up to 10-5 W/m^2 during flare eruptions. A period of 2:55UTC to 3:05UTC marked the beginning of the flare with a value of M1.7. Minutes later, CME was detected using LASCO C3 coronagraph in figure 2, where radiation particles erupted in the solar atmosphere. Space weather forecasts indicate the eruption occurred near N16W14, which is slightly facing Earth. There is a strong probability of a geomagnetic storm based on the location of the flare [10].

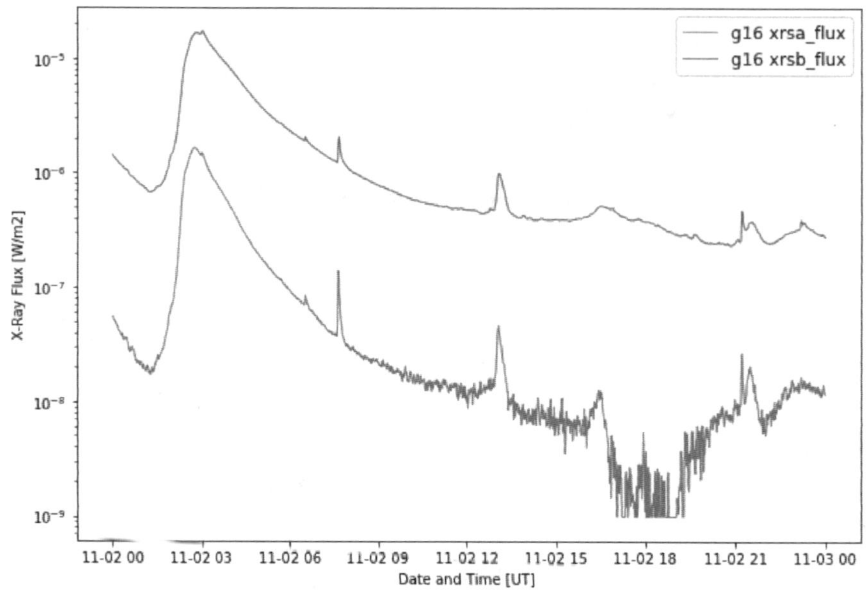

Figure 1. X-ray flux during 2nd November 2021.

Figure 2. Coronagraph image from LASCO C3 at 2nd November 2021.

3.2 GOES-16 magnetometer (MAG)

Space weather activities are predicted and determined by observations of the magnetic field. Geomagnetic storms are detected by GOES-R by measuring the magnitude and direction of the Earth's magnetic field. An EPN coordinate system is used for MAG data, in which E is the Earthward (nadir) direction, N is the eastward velocity direction, and the P-axis is roughly parallel to the Earth's spin axis[7]. The magnetic field was distributed during the flare on 2nd November in figure 3a, indicating that there was a disturbance in the magnetic field in the surrounding environment. In figure 3b, the magnetic field value appeared to be quiet at the start of 3rd November, but at about 19:30 UTC, there was a sharp decrease. In figure 3c, a more severe component of the interplanetary magnetic field rapidly decreased

Journal of Physics: Conference Series **2543** (2023) 012011 doi:10.1088/1742-6596/2543/1/012011

to as low as -50nT. This plasma collision within the magnetosphere caused a moderate G3 geomagnetic storm conditions that allowed for aurora visibility as seen in figure 5 from NOAA aurora forecast.

3.3 GOES-16 Space Environment In-Situ Suite (SEISS)

GOES-16 measures proton flux in different energy channels. We plotted the mean energy differential proton flux using GOES-R level2 data and the Python program. Making it easier to analyse its peak time and detects the value of proton flux within the atmosphere. Figure 4 shows the average differential proton flux (AvgDiffProtonFlux) for 2nd-4th November 2021. On 2nd November, the AvgDiffProton. Flux was within average of 10^2-10^4 protons/cm2-s-sr, but around 3:00 UTC there was a sharp increase that might have been triggered by the flare and CME explosion. Figure 4b shows that at the start of the 3rd of November, the AvgDiffProtonFlux was in an average state, but there was a huge increase in its value. This might be an indication of high charged particle concentrations in the magnetosphere. In Figure 4c, the AvgDiffProtonFlux was still at a high value but wasn't exceeding 10^5 protons/cm2-s-sr, the spike continued within the range of 10^5-10^4 protons/cm2-s-sr. This also shows the possibility that a geomagnetic storm was still active within the two days right after the flare eruption of AR12891.

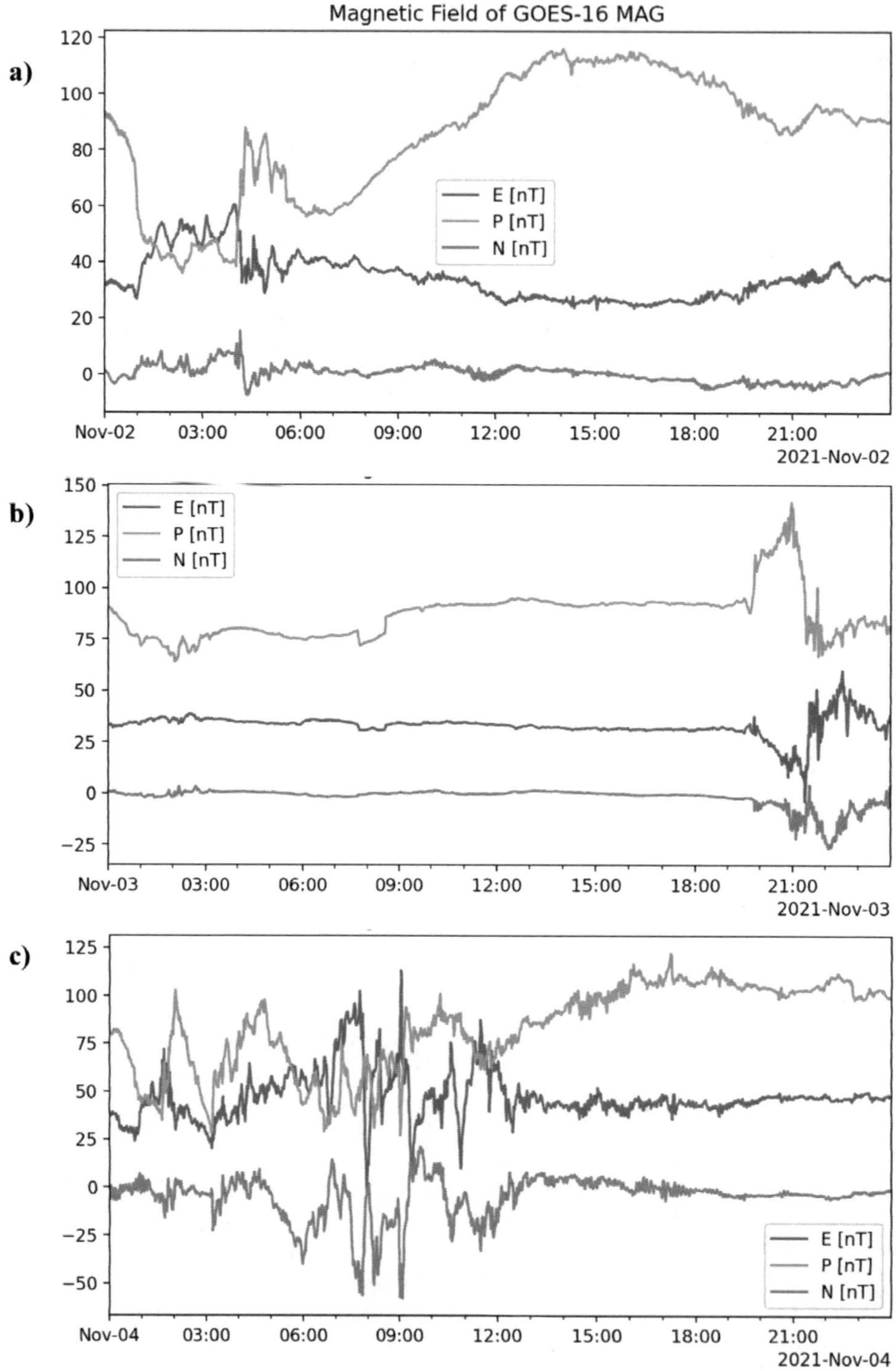

Figure 3. GOES-16 MAG measuring magnetic field in space environment during a) 2nd November 2021 b) 3rd November 2021 and c) 4th November 2021.

a)
b)
c)

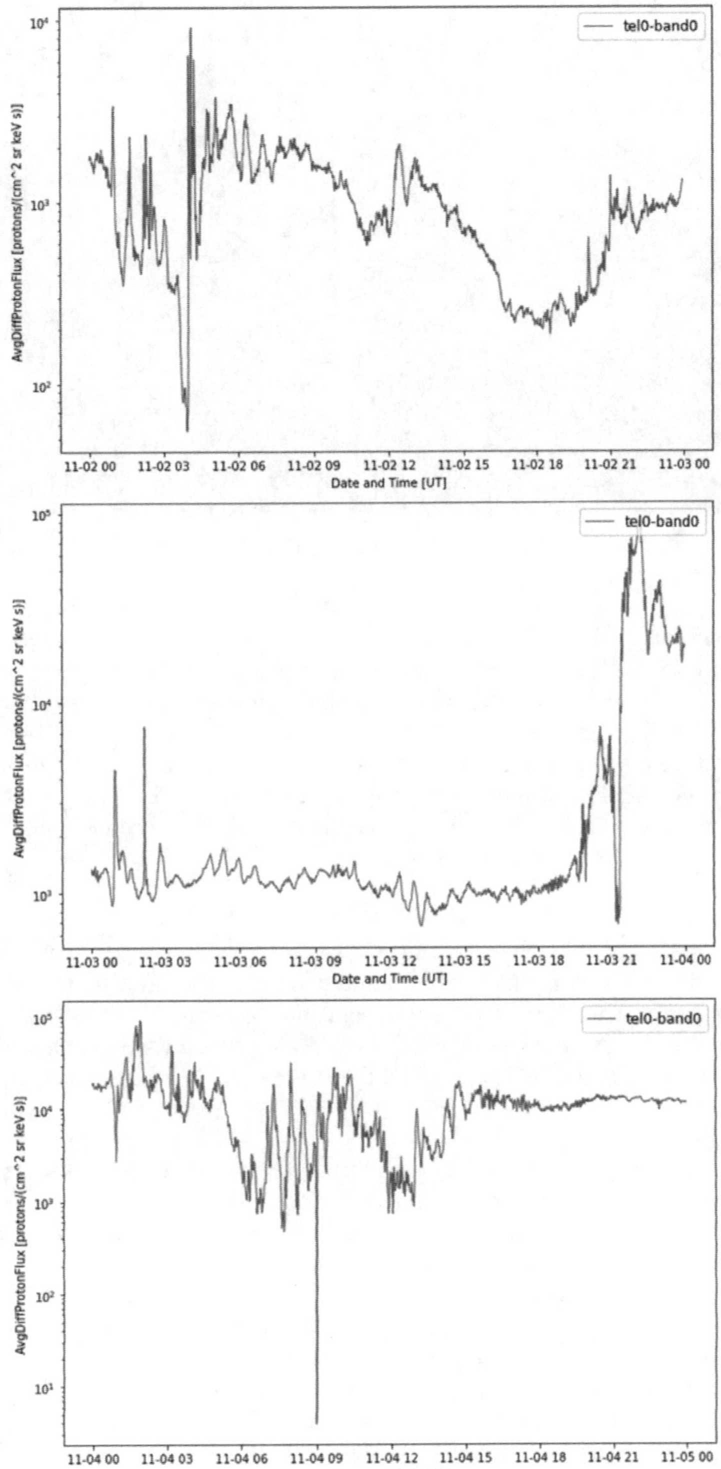

Figure 4. GOES-16 SEISS measuring average proton flux during a) 2nd November 2021 b)3rd November 2021 and c) 4th November 2021.

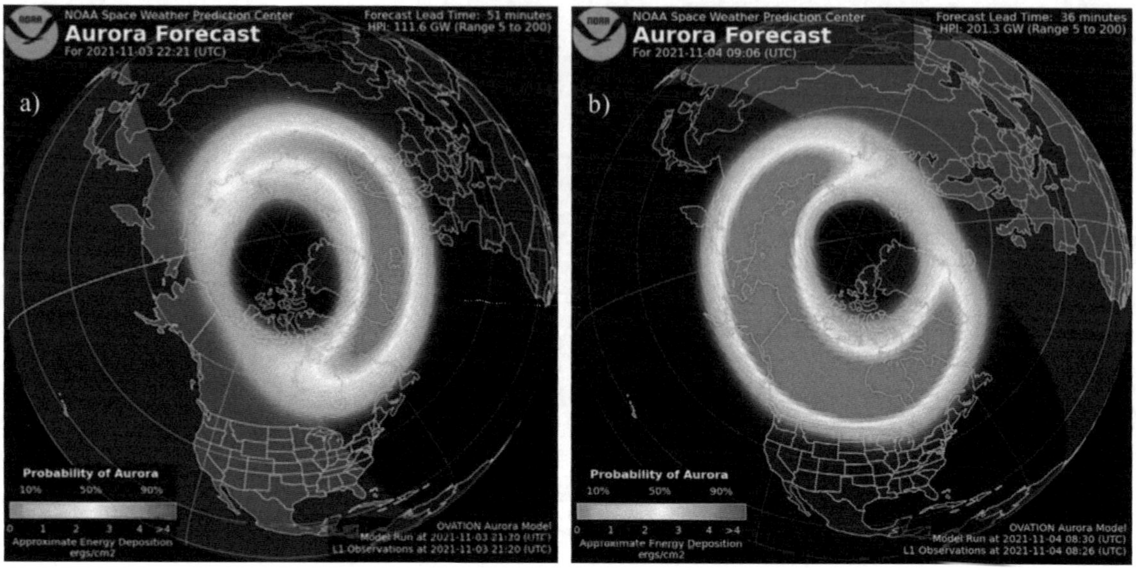

Figure 5. NOAA Aurora forecast after the M1.7 flare eruption.

4. Conclusion

On 4th November 2021, the strongest flare on 2nd November 2021 resulted in a moderate G3 geomagnetic storm. In 2021, AR 12891 produced the highest geomagnetic storm of solar cycle 25. By watching the flare's growth, NOAA could detect any solar storms and avoid any satellite communication loss. The GOES-16 equipment can be used to measure the x-ray flux, proton flux, and magnetic field plasma that would disrupt the earth's magnetosphere during a flare. Through the use of GOES-16 instruments, space weather forecasting has been greatly improved. It is our hope that GOES-16 data will be improved in any way, whether through enhancement of existing products, the development of new products, or the correction of known problems.

Acknowledgements

Special gratitude to SDO/AIA, NOAA, SWPC, LASCO, Helioviewer, Solar Monitor, and Space Weather Live websites for making their data available to the public. The authors also show their respect to the reviewers and editors for their insightful remarks and recommendations. This project under Young Talent Research Grant 600-RMC/YTR/5/3 (001/2020) and FRGS Grant 600-IRMI/FRGS 5/3 (309/2019) Kementerian Pengajian Tinggi (KPT)

References

[1] Jilani S A and Wajih S 2018 Correlation Structure of Solar Radio Bursts in Accordance with Solar Activity *Journal of Basic & Applied Sciences* **14** 17–22

[2] Melrose D B 2017 Current-driven flare and CME models *J Geophys Res Space Phys* **122** 7963–78

[3] Piddington J H *GEOMAGNETIC STORMS, AURORAS AND ASSOCIATED EFFECTS*

[4] Kress B T, Rodriguez J v. and Onsager T G 2019 The GOES-R Space Environment In Situ Suite (SEISS): Measurement of Energetic Particles in Geospace *The GOES-R Series: A New Generation of Geostationary Environmental Satellites* (Elsevier) pp 243–50

[5] Machol J L, Eparvier F G, Viereck R A, Woodraska D L, Snow M, Thiemann E, Woods T N, McClintock W E, Mueller S, Eden T D, Meisner R, Codrescu S, Bouwer S D and Reinard A A 2019 *GOES-R Series Solar X-ray and Ultraviolet Irradiance* (Elsevier Inc.)

[6] Redmon R J, Seaton D B, Steenburgh R, He J and Rodriguez J v. 2018 September 2017's Geoeffective Space Weather and Impacts to Caribbean Radio Communications During Hurricane Response *Space Weather* **16** 1190–201

[7] Loto'aniu T M, Redmon R J, Califf S, Singer H J, Rowland W, Macintyre S, Chastain C, Dence R, Bailey R, Shoemaker E, Rich F J, Chu D, Early D, Kronenwetter J and Todirita M 2019 The GOES-16 Spacecraft Science Magnetometer *Space Sci Rev* **215**

[8] Loto'aniu P T M, Califf S, Redmon R J and Singer H J 2019 Magnetic Field Observations from the GOES-R Series *The GOES-R Series: A New Generation of Geostationary Environmental Satellites* (Elsevier) pp 251–9

[9] Sphinx 2021 goesr-spwx-examples documentation

[10] Rossdee A and Shariff N N M 2022 Observing the Development of Two Solar Flares by Monitoring AR 12882 which Produces a Kp6 Geomagnetic Storm in October 2021 *J Phys Conf Ser* **2287** 012032

Atomic hydrogen adsorption on freestanding α-borophene nanoribbons for hydrogen spillover applications: Insights from density functional theory calculations

A B Santos-Putungan*, M U Herrera and D B Putungan

Institute of Mathematical Sciences and Physics, University of the Philippines Los Baños, Philippines

* Corresponding author's email: absantos1@up.edu.ph

Abstract. In this work, we probed the applicability of α-borophene nanoribbon as a spillover hydrogen host material using density functional theory (DFT) calculations. It is found that all top sites, along the edges and on the planar surface, are appropriate for hydrogen physisorption. On the other hand, the hollow site which is a defining feature of the α-borophene nanoribbon, was found to be an unstable adsorption site for hydrogen, and does not act as a trap site for hydrogen atom. Density of states calculations show that the broadening of the hydrogen 1*s* states made it possible for a good hybridization with the boron 2*p* states, allowing for such a moderate hydrogen physisorption. The resulting moderate adsorption energies indicate good hydrogen migration properties, making α-borophene nanoribbon a very good candidate material for spillover hydrogen applications and could be utilized as a key material for overall hydrogen storage.

1. Introduction

Due to the disadvantages and harm brought by the use of fossil fuels to our environment, the search for an alternative and cleaner energy resource has gained wide attention in recent years. Among potential alternatives, hydrogen emerges as one of the most viable candidates as an energy carrier, as it does not emit greenhouse gases, and has one of the highest energy density per mass [1]-[2]. In principle, the earth acts as an unlimited source of hydrogen due to the fact that it consists of more than 70% water, assuming that water splitting could proceed via a sustainable pathway. Hydrogen gas can then be produced via hydrogen evolution reaction (HER) via cheap and abundant catalysts [3]-[7].

In order for hydrogen to be utilized efficiently, appropriate storage materials must be identified. Solid state hydrogen storage materials such as Li-decorated single layer MoS_2 has been shown to effectively store molecular hydrogen, especially at full Li coverage [8]-[9]. Here, it was also demonstrated that molecular hydrogen can be dissociated, allowing for the spillover mechanism to proceed. As such, the Li-decorated single layer MoS_2 can be effectively employed as a molecular hydrogen storage and at the same time can be used to dissociate hydrogen. The dissociated hydrogen atoms can then diffuse to another material to be adsorbed. Therefore, both molecular and atomic hydrogen storage play an important role in the overall hydrogen storage process.

Most hydrogen spillover studies focus on graphene and carbon nanotubes [10]–[15]. In light of the popularity of graphene, several other 2D materials such as transition metal dichalcogenides (TMDs) have been shown to possess certain catalytic properties that could also be tapped for heterogeneous catalysis applications [16]–[18]. Recently, borophene emerged as one of the most extensively studied materials due to its excellent properties similar to that of graphene [19]–[22]. Different allotropes of 2D-borophene sheets have been investigated including α-, β-, χ-, and δ-type nanostructures [23]–[28]. In addition, 2D-borophene sheets and nanoribbons were found to exhibit good electrical, chemical, and mechanical properties [29]–[33]. In this work, we looked into the applicability of semi-infinite borophene nanoribbons as storage materials for atomic hydrogen as a result of spillover process from a molecular hydrogen host material, via density functional theory calculations (DFT).

2. Methodology

All calculations were performed using the Quantum Espresso simulation package [34]. Spin-polarized, dipole-corrected DFT calculations utilizing the Perdew-Burke-Ernzerhof (PBE) parametrization of the Generalized Gradient Approximation (GGA) as the exchange-correlation functional [35] were used. Vanderbilt ultra-soft pseudopotentials were used for both the B and H atoms. Atomic forces and electronic convergence were set to 0.001 a.u and 0.00001 a.u, and cut-off energy of 60 Ry were used for structure optimization. A 15 Å vacuum was used to avoid the structure interactions. The Monkhorst-Pack *k*-point mesh was set at 6 × 1 × 1 to sample the **k**-space.The semi-infinite α- borophene structure was constructed based from the work of [36]. The structure considered is 5.93 A in length from its finite size (y-axis).

Adsorption energy was computed using

$$E_{ads} = E_{B-H} - E_B - \frac{1}{2}E_{H_2}$$

where E_{ads} is the adsorbed energy of H on borophene, E_{B-H} is the total energy of borophene with hydrogen, E_B is the energy of borophene, and E_{H2} is the energy of Hydrogen molecule. These energies were directly calculated via DFT calculations using the parameters mentioned above. In the above definition, the more negative the adsorption energy value is, the more stable is the adsorption of H on the B sheet.

3. Results and discussion

Figure 1 shows the different sites on α-borophene nanoribbon wherein hydrogen atom physisorption was found. These sites are top sites of α- borophene nanoribbon along its infinite axis edge (S1 and S2), along the finite axis edge (S3) and along the perimeter of the hollow site within the nanoribbon's plane (S4).

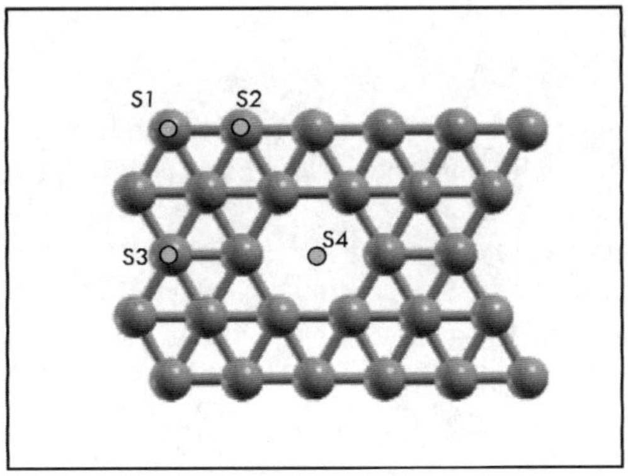

Figure 1. Sites of H atom (in pink) where adsorption is observed. The H atoms are placed on top of borophene before optimization.

Listed in table 1 are the calculated adsorption energies of the hydrogen atom on the considered sites. It can be observed that there are stable physisorption of hydrogen atom on all sites. Magnitude wise, the adsorption values are effectively the same on all sites, with those on S3 and S4 sites slightly higher compared to that of S1 and S2. This implies that for the α-borophene nanoribbon, all top sites whether along the edges or within the plane are potential binding sites for hydrogen atom. On a note, we considered adsorbing hydrogen at the hollow site center and upon relaxation, the hydrogen atom moved to a nearby S4 site. This shows that for the α-borophene nanoribbon, the hollow site which is essentially a defect on the planar surface, is not conducive to atomic hydrogen adsorption and does not act as a trap site, a common feature for most in-plane defects. Shown in figure 2 are the adsorption geometries for the two most stable adsorption sites (S3 and S4). These moderate physisorption values are appropriate for spillover hydrogen applications, as hydrogen atoms are not tightly bound on the surface of α-borophene nanoribbon. This will allow easy migration of hydrogen atoms onto other sites or onto other related spillover hosts.

Table 1. Calculated adsorption energies of H-atom where an adsorption was found.

Adsorption Sites	E_{ads} (eV)
S1	-0.24
S2	-0.24
S3	-0.27
S4	-0.28

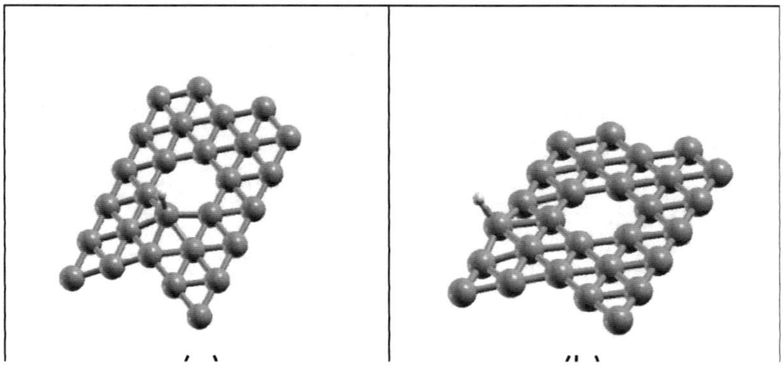

Figure 2. Optimized structure where adsorption of H atom can be observed. H atom is seen to have chemisorption with borophene on sites near the edges and defect.

Figure 3 shows the atom-projected density of states plot for the representative adsorption site S4. It is noticeable that the hydrogen 1*s* states have broadened and hybridized well with the 2*p* states of boron. Such hybridization resulted to moderate physisorption on the considered binding site. This property of the α-borophene nanoribbon with regard to hydrogen adsorption makes it a potential host for spillover hydrogen applications, and could be coupled to hydrogen storage systems for a sustainable overall hydrogen storage mechanism.

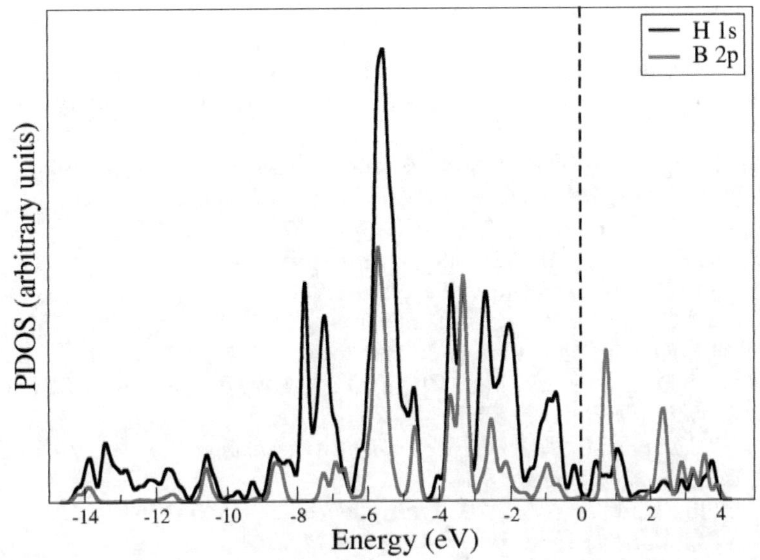

Figure 3. Projected density of states (PDOS) plot for the representative adsorption site S4 showing the hybridization of 1s state of hydrogen with the 2p state of boron.

4. Conclusion

Density functional theory calculations were carried out to investigate the potential of α-borophene nanoribbon as a hydrogen spillover material. It is found that all top sites along the infinite and finite axes edges, as well as on the surface plane, are stable adsorption sites for hydrogen physisorption. The hollow site on the other hand, is found to be unstable for atomic hydrogen adsorption, implying that such defect-like structure does not act as a trap site for hydrogen atom. Density of states showed that the good hybridization between hydrogen 1*s* and boron 2*p* states resulted to moderate hydrogen physisorption that allows for easy migration of hydrogen atoms on the surface and as such, making the hydrogen spillover mechanism at work. The results of this work are of significance in the further elucidation of hydrogen adsorption properties on low-dimensional boron materials, and could be potentially tapped for overall hydrogen storage materials.

Acknowledgement

This work was funded by the University of the Philippines Los Baños Basic Research Grant 2020. AB Santos-Putungan would like to acknowledge the Computing and Archiving Research Environment (COARE) of DOST - Advanced Science and Technology Institute, Philippines for providing the computational facilities.

References

[1] Dunn S 2002 *International Journal of Hydrogen Energy* **7** 3
[2] Shinnar R 2003 *Technology in Society*. **25** 4
[3] Pu, Z *et al.*, *Nano-Micro Letters* **12** 1
[4] Olbigacion, J K A and Putungan, D B 2020 *Mater Res Express* **7** 7
[5] Wu M H, Chou W J, Huang J S, Putungan D B and Lin S H 2019 *Physical Chemistry Chemical Physics* **21** 38
[6] Zhong W, *et al.* 2021 *Advanced Materials* **33** 9
[7] Putungan D B, Lin S H and Kuo J L 2015 *Physical Chemistry Chemical Physics* **17** 33
[8] Putungan D B, Lin S H, Wei C M and Kuo J L 2015 *Physical Chemistry Chemical Physics* **17** 17
[9] Wu L, *et al.* 2019 *ACS Energy Lett* **4** 7
[10] Yodsin N, Sakagami H, Udagawa T, Ishimoto T, Jungsuttiwong S and Tachikawa M 2021 *Molecular Catalysis* **504**

[11] Guo J H, Liu J X, Wang H B, Liu H Y and Chen G, 2021 *Physical Chemistry Chemical Physics* **23** 3
[12] Guo J H, Li X D, Cheng X L, Liu H Y, Li S J and Chen G 2018 *Int J Hydrogen Energy* **43** 41
[13] Sha Z, Knippenberg M T, Cooper A C, Pez G P and Cheng H 2008 *Journal of Physical Chemistry C* **112** 44
[14] Chen C H and Huang C C 2008 *Microporous and Mesoporous Materials* **109** 1–3
[15] Chen L, Cooper A C, Pez G P and Cheng H 2007 *Journal of Physical Chemistry C* **111** 51
[16] Santos-Putungan A B and Putungan D B 2019 *Mater Res Express* **6** 6
[17] Sevilla J R M and Putungan D B 2021 *Mater Res Express* **8** 8
[18] Putungan D B and Lin S H 2017 *Mater. Res. Express* **4**
[19] Kong L, Wu K and Chen L 2018 *Frontiers of Physics* **13** 3
[20] Ranjan P, Lee J M, Kumar P and Vinu A 2020 *Advanced Materials* **32** 34
[21] Yin Y, Li D, Hu Y, Ding G, Zhou H and Zhang G 2020 *Nanotechnology* **31** 31
[22] Li W, *et al.* 2018 *Sci Bull (Beijing)* **63** 5
[23] Vishkayi S I and Tagani M B 2018 *Physical Chemistry Chemical Physics* **20** 15
[24] Ou M *et al.* 2021 *Advanced Science* **8** 12
[25] Taşaltın N, Güllülü S and Karakuş S 2022 *Inorg Chem Commun.* **136**
[26] Duo Y, *et al.* 2021 *Coordination Chemistry Reviews* **427**
[27] Mozvashi S M, Mohebpour M A, Vishkayi S I and Tagani M B 2021 *Sci Rep* **11** 1
[28] He J, Ouyang Y, Yu C, Jiang P, Ren W and Chen J 2020 *Chinese Physics B* **29**
[29] Feng B *et al.*2016 *Nat Chem* **8** 6
[30] Mortazavi B, Rahaman O, Dianat A and Rabczuk T 2016 *Physical Chemistry Chemical Physics* **18** 39
[31] Giannopoulos G I 2017 *Comput Mater Sci* ***129***
[32] Gao Z Y, Wang W, Sun L, Yang L M, Ma B Y and Li P S 2022 *J Magn Magn Mater* **548**
[33] Mannix A J, Zhang Z, Guisinger N P, Yakobson B I and Hersam M C 2018 *Nat Nanotechnol*, **13** 6
[34] Giannozzi P *et al.* 2009 *J. Phys.:Condens. Matter* **21**
[35] Perdew J P, Burke K and Ernzerhof M 1997 *Phys. Rev. Lett.* **77** 3865
[36] Tang H and Ismail-Beigi S 2007 *Phys Rev Lett* **99** 11

An analysis of slow drift type II radio burst and its association with full halo coronal mass ejection on 12th January 2022

SNA Shamsuddin[1], NM Ansor[1], ZS Hamidi[1,2] and NNM Shariff[2,3,*]

1 Faculty of Applied Sciences, Universiti Teknologi MARA Shah Alam, 40450 Shah Alam, Selangor, Malaysia
2 Institute of Science, Universiti Teknologi MARA Shah Alam, 40450 Shah Alam, Selangor, Malaysia
3 Academy of Contemporary Islamic Studies (ACIS), Universiti Teknologi MARA Shah Alam, 40450 Shah Alam, Selangor, Malaysia

* Corresponding author's e-mail: nnmsza@uitm.edu.my

Abstract. Radio bursts have been a point of contention among astrophysicists in the modern era. Solar flares, coronal mass ejections (CMEs), and solar energetic particles (SEPs) that project from the Sun to interplanetary space can all be observed using this method. Based on data from e-Callisto, we detected a type II radio burst on 12th January 2022 starting at 04:24 UT and lasting until 04:56 UT. The full image of radio burst type II (SRBTII) was compiled from two stations (Almaty and Australia-ASSA), the drift rate of the bursts calculated are -0.07796 MHz/s and -0.105822893 MHz/s respectively, and the relevant solar activity was determined. There was a possible association between the observed burst and the halo CME event on the day. This study includes the properties of the CME and the D-region Absorption Prediction (D-Rap) plot as the signal degradation indicator caused by the solar flare and CMEs to demonstrate its significance in understanding the impact on Earth's atmosphere.

1. Introduction

The solar flares and coronal mass ejections (CMEs) that originate from our Sun are believed to have a significant impact on climates and conditions in space. Observable waves and radiation are emitted by these activities. Since 1985, the theory on radio wave diagnostics of the solar processes at meter and decimetre has been started by McLean and Labrum and continued by Benz in 2002[1]. Observations of solar activity via radio waves have proven to be one of the most effective methods of solar monitoring. Due to the frequency range of radio, the particle's sudden acceleration can be visualized. Typically, five types of bursts can be classified to Type I (noise storm), Type II (slow drift rate), Type III (fast drift rate), Type IV, and Type V. [2, 3]. However, Type II, III, and IV are usually most relevant to space weather conditions[4]. Figure 1 shows the schematic diagram of the burst on the frequency-time graph. A variation in the frequency over time determines the type of burst.

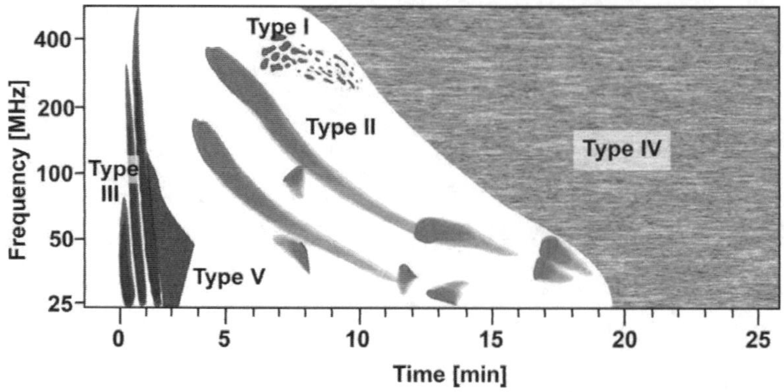

Figure 1. Schematic diagram of Radio burst type I, II, III, IV, and V on frequency-time plot graph.

CALLISTO stands for Compound Astronomical Low-Frequency Low-Cost Instrument Spectroscopy is a programmable heterodyne receiver designed by Christian Monstein, a member of the former Radio Astronomy Group (RAG) in 2006 at ETH Zurich, Switzerland. This instrument operates between 45 and 870 MHz[5]. These days, the CALLISTO spectrometer has been widely used in many countries which makes it able to observe the sun's activity 24 hours per day. eCALLISTO's websites will automatically upload data from all stations, and images of the observations will be publicly available.

Type II Burst (SRBTII) and Coronal Mass Ejections (CMEs)

Type II bursts were first discovered as radio emission that progressively starts at low frequency before being classified as type II bursts in 1957 by Wild and McCready by Payne-Scott, Yabsley, and Bolton in 1947[6]. Solar burst type II (SRBTII) is said to be associated with CMEs as it is excited by the magnetohydrodynamic (MHD) shockwaves and usually occurs at the peak time of the solar flare. It is a class of intense emission that is prone to propagate shockwave close to the electron plasma frequency, fp or near 2fp, and proportionally to the square root of electron number density, n_e[7, 8]. The burst shift from higher to lower frequency over a long time compared to Type III (SRBTIII), therefore it is classified as a slow drift burst. The drift rate of the burst can be calculated by using the equation below;

$$Drift\ rate = \frac{df}{dt} = \frac{f_e - f_s}{t_e - t_s}\left(\frac{MHz}{s}\right) \tag{1}$$

where f_e is the frequency of end time, f_s is the frequency of start time, t_e is the end time, and t_s start time. This power law was also valid when observations from different spectral domains [metric (m), decameter-hectometric (DH), and kilometric (km) wavelengths] are combined for the same events or different events.

Coronal Mass Ejections (CMEs) are one of the massive events that emit radiation and highly energetic mass particles from the Sun's corona to interplanetary space. The speed of the CMEs can reach up to 300 km/s while the mass can be up to 10^{13}kg. It has been proven that the shock waves from CMEs consist of large and spatial varying magnetic fields which lead to intensively fast plasma flows [9, 10]. As a result of this massive event, the Earth's magnetic field may be disrupted and space weather changes, such as geomagnetic storms, may occur. An induced high voltage from a geomagnetic storm could damage the electrical power grid. A CME can also alter the ionospheric condition, damage the spacecraft, and degrade data due to energetic particles. To avoid such events, it is crucial to know how to take precautions. SRBTII provides information about the properties of the event by indicating the emission excited by CMEs. This paper discusses the properties of SRBTII from the CALLISTO spectrometer and the relevant CMEs.

2. Methodology

A Compound Astronomical Low-cost Instrument for Spectroscopy and Transportable Observatory (CALLISTO) data were used for this paper. CALLISTO spectrometer burst data from all stations are compiled in this open-access database. Currently, there are more than 150 instruments that have been implemented all over the world. The parameters that are involved in this analysis are emission frequencies, time, and duration of the bursts which are extracted by RAPPVIEWER to obtain the accurate reading of frequencies at a specific time. Meanwhile, the properties and images of the full halo CME are obtained from SOHO LASCO Catalogue before both events are analysed and correlated. Our next step is to correlate the events with the CMEs recorded by LASCO. A list of CME events has been provided by LASCO along with parameters such as velocity, angular width, and mass. To analyse the emission frequencies, time, and duration of bursts, we extract the data using RAPPVIEWER software which gives us the exact frequency at the certain time selected. Additionally, D-region Absorption Prediction (D-Rap) from NOAA provides real-time global estimates on (i) Polar Cap Absorption (PCA) due to SEPs and the X-ray absorption due to solar x-ray flares. Highlighted part on the images allows us to see the intensity and the area of locations affected due to the disturbances within the ionospheric D-region which are mainly used for high-frequency (HF) communications.

3. Observations

On 12th January 2022, a type II radio burst was recorded by the CALLISTO system and the burst lasted for 32 minutes from 04:24 UTC to 04:56 UTC. ALMATY station which is in Kazakhstan shows a brief burst from 04:26 UTC to 04:31 UTC (figure 2). Apart from ALMATY, Australia-ASSA station also recorded the same type of bursts which began 2 minutes before Almaty station and eased off at 4:50 UT.

4. Discussion

As we combined these two bursts from two different stations that cover different ranges of frequency, it convinced us that this burst varied from ~20 MHz to ~140 MHz. Based on figure 2, we can see that Australia-ASSA detected the burst longer than ALMATY. ALMATY station has some unknown disturbances below 60 MHz, which is why the burst was detected heavily at frequencies lower than 60 MHz. The burst image has also become clearer at Australia-ASSA due to the larger scale of the frequency.

The white lines in figure 2 show the frequency range of each fragment to determine the maximum and the minimum frequency and calculate the drift rate of the bursts. By using equation (1), the average drift rate of both observations was calculated based on the pattern of white line on the figure 2 and tabulated in table 1. The value of the drift rate agreed with the ranges stated by previous studies on SRBTII, the slow drift burst with typically ≤50MHz/s at metric wavelengths and would appear at lower latitude of frequency over longer wavelength [11, 12].

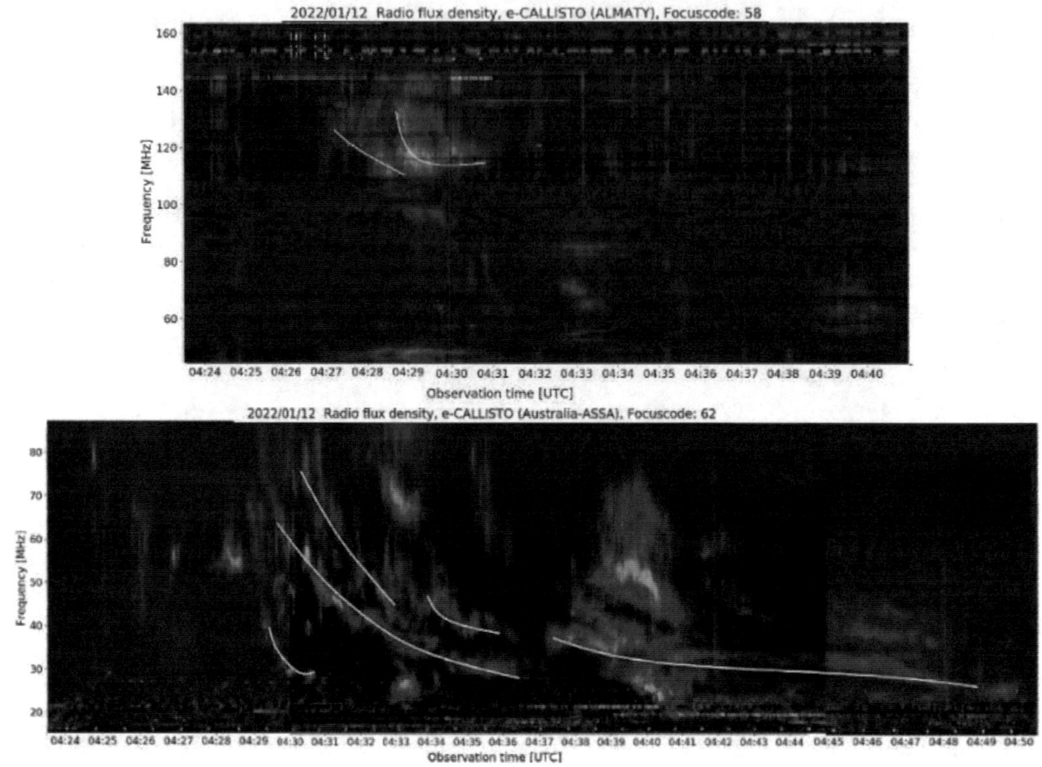

Figure 2. Time-Frequency graph of type II burst from (a) ALMATY (60-160 MHz) and (b) Australia-ASSA (20-80 MHz) with frequency range (white line) for drift rate calculation (Retrieved from: https://www.e-callisto.org/index.html)

Table 1. Average drift rate from ALMATY and Australia-ASSA station.

Station	Average drift rate
ALMATY	-0.07796 MHz/s
Australia-ASSA	-0.105822893 MHz/s

On the same day, a full halo CME observed by the LASCO C2 coronagraph emerged from the northeast part of the solar limb (figure 3). The eruption first appears in the C2 image at 04:36 UT and dominated the east side of the graph. In the second and third frames, the CME is actively evolved, producing a full halo CME which propagated at a speed of 1586 km/s, accelerating at -40.8 m/s^2 at 360 angular widths. The shock wave generated by this CME caused the speed of the solar wind to increase to 402.4 km/sec and the density reported on the day was 6.6 $proton/cm^3$. On the event day, there were only B-class flares reported on 12th January 2022.

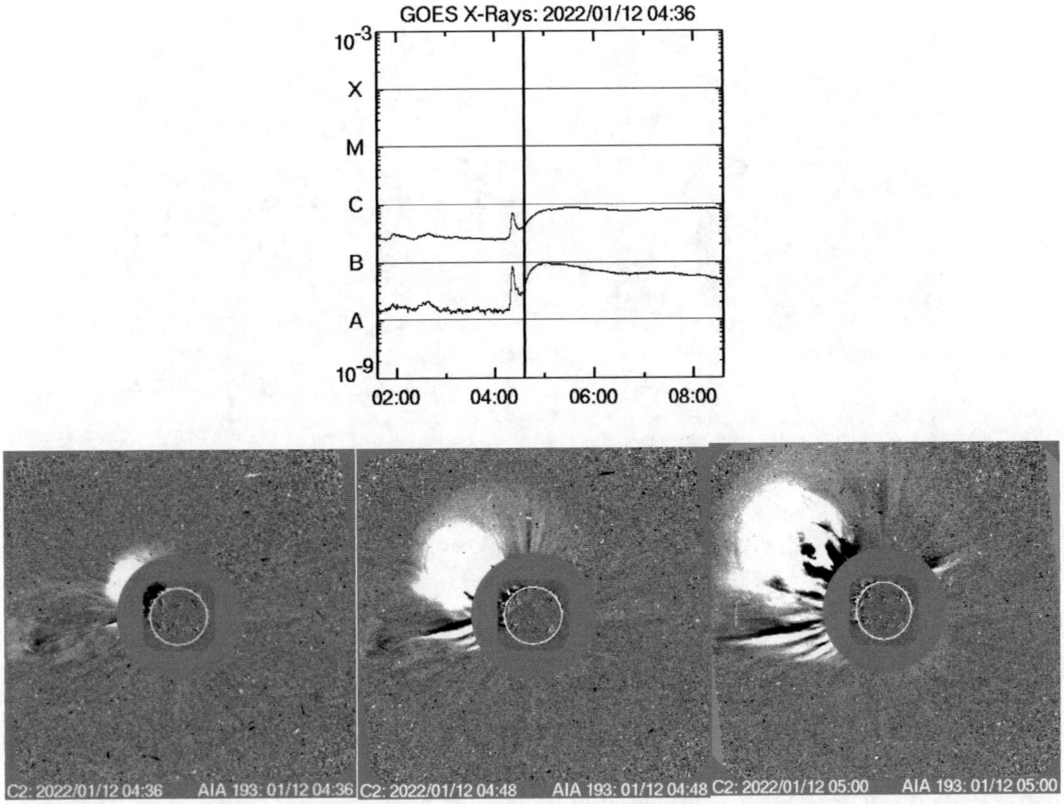

Figure 3. (i) The upper panel shows the X-ray flux density measured by the GOES X-Rays instrument during the expulsion happen. The solid line indicates the x-ray flux during the CME event starting. (ii) The lower panel illustrates the evolution of expulsion which starts at 04:36 UT and evolved producing a full halo CME. The bright structure of the emission reflecting the expulsion has relatively increasing intensity. (Credit to: LASCO Catalog)

Following the CME event, we can see some increments in degraded frequency in the D-Rap plot for the next half an hour (figure 4). We assumed it was due to energetic protons released by CME triggering an increment of the free electron within the ionosphere and enhancing the radio wave absorption. However, it was not significantly affected and normal x-ray and proton in the background reported as there was no high-class flare reported. Therefore, there is not much signal attenuation due to this event. This assumed to happen due to the initial direction of the eruption which needs to be deeper research on it in the future.

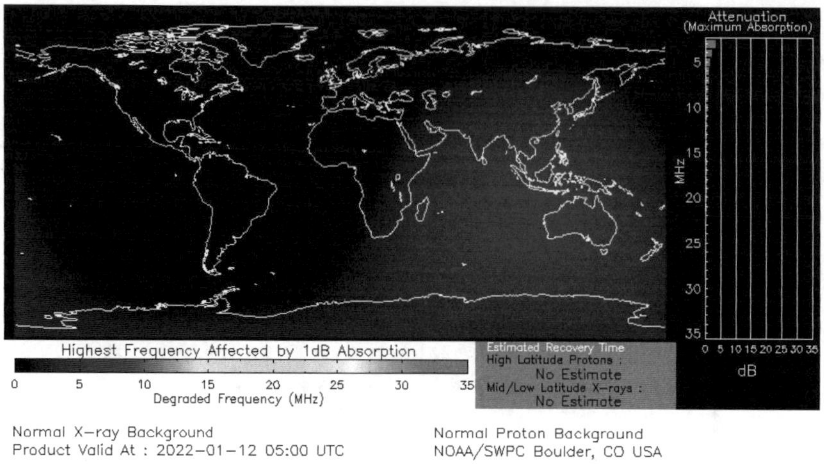

Figure 4. D-rap plot graph that indicates the degraded frequency affected due to solar flares, coronal mass ejections, and solar energetic particle events. (Retrieved from https://www.ngdc.noaa.gov/stp/drap/index.html)

5. Conclusion

The type II radio burst started at 04:24 UT with 80 MHz recorded by ALMATY and Australia-ASSA. The average drift rates obtained by Almaty and Australia-ASSA are -0.07796 MHz/s and -0.105822893 MHz/s respectively. It also known as slow drift bursts, exhibit these characteristics. A CME event occurs shortly after the burst appears on the northeast limb side of the sun from LASCO C2. Due to the absence of solar flares on that day, type II radio bursts can mostly be attributed to a CME that was induced by the magnetohydrodynamic shock that caused the eruption. Furthermore, this event is not intense enough to give a significant impact on the ionospheric conditions and cause shortwave fadeout. This study shows that, although radio observations are usually associated with solar flares, they can also be used to observe CMEs. Besides, the continuation of studies on flares like this will give evidence on relating radio bursts with CME, one of the major solar events.

Acknowledgement

We thanks to CALLISTO network, LASCO, STEREO, SDO/AIA, NOAA and SWPC for make their data available online. We are grateful to NOAA Space Weather Prediction Centre (SWPC) for sunspot, radio flux and solar flare data for comparison purpose. The authors also thank the reviewers and editors for their valuable comments and suggestions. We acknowledge the support provided by grants 600-RMI/FRGS 5/3 (309/2019), 600-RMC/YTR/5/3 (001/2020) and Kementerian Pengajian Tinggi Malaysia.

References

[1] Benz A, Monstein C, Meyer H, Manoharan P, Ramesh R, Altyntsev A, et al. A world-wide net of solar radio spectrometers: e-CALLISTO. Earth, Moon, and Planets. 2009;104(1):277-85.

[2] Behlke R. Solar radio bursts and low frequency radio emissions from space. IRF Scientific Report 275. 2001.

[3] Wijesekera J, Jayaratne K, Adassuriya J, editors. Analysis of type II and type III solar radio bursts. Journal of Physics: Conference Series; 2018: IOP Publishing.

[4] Ramli N, Hamidi Z, Abidin Z, Shahar S, editors. The relation between solar radio burst types II, III and IV due to solar activities. 2015 International Conference on Space Science and Communication (IconSpace); 2015: IEEE.

[5] Monstein C. e-Callisto [Available from: https://www.e-callisto.org/index.html.]

[6] Gopalswamy N, Mäkelä P, Yashiro S. A catalog of type II radio bursts observed by Wind/WAVES and their statistical properties. arXiv preprint arXiv:191207370. 2019.

[7] Schmidt JM, Cairns IH, Lobzin VV. The solar type II radio bursts of 7 March 2012: Detailed simulation analyses. Journal of Geophysical Research: Space Physics. 2014;119(8):6042-61.
[8] Palmer ID. A type II solar radio burst observed in the corona and in interplanetary space. Solar Physics. 1974;37(2):443-9.
[9] St. Cyr O, Howard R, Sheeley Jr N, Plunkett S, Michels D, Paswaters S, et al. Properties of coronal mass ejections: SOHO LASCO observations from January 1996 to June 1998. Journal of Geophysical Research: Space Physics. 2000;105(A8):18169-85.
[10] Gopalswamy N. Coronal mass ejections of solar cycle 23. Journal of Astrophysics and Astronomy. 2006;27(2):243-54.
[11] McLean DJ, Labrum NR. Solar radiophysics: Studies of emission from the Sun at metre wavelengths. United States: Cambridge University Press; 1985.
[12] Gopalswamy N. Coronal Mass Ejections and Type II Radio Bursts. Solar Eruptions and Energetic Particles2006. p. 207-20.

Power-frequency performance of step-graded double Gunn diodes

L J Butts*, S M El-Ghazaly

Department of Electrical Engineering, University of Arkansas, Fayetteville, AR, 72701, USA

* Corresponding author's e-mail: Lb031@uark.edu

Abstract. Simulations of conventional Gunn diodes containing the double Gunn effect have shown a significant increase in oscillation frequency and frequency range. However, conventional Gunn diodes suffer from higher phase noise, lower efficiency, and reduced temperature stability due to the lack of a hot electron injector. A study on the performance of a Double Gunn diode with a graded gap injector is presented. The second region of negative differential mobility is gradually introduced in several temperature-dependent mobility models, which are implemented in an energy balance simulation. Devices with transit lengths of 0.7μm, 1.1μm, and 1.65μm are simulated. All devices experience an increase in oscillation frequency with increasing intensities of the double Gunn effect. The frequency range doubles at significant intensities of the effect in the devices with transit lengths of 0.7μm and 1.65μm.

1. Introduction

The development of high-performance millimeter wave oscillators continues to be a priority for the transportation, defense, and communication industries since they construct detection, measurement, and imaging systems that advance these industries. While efficiency and power performance metrics are more relevant to certain applications, low noise and wide bandwidth oscillators continue to be the enabling technology for many applications. The Gunn diode has been reliably used for the low-noise generation of millimeter waves with moderate output power. However, the electronically tunable frequency range for these devices is limited.

The double Gunn diode was previously presented as a solution for generating low-noise, ultra-wide bandwidth mm-wave signals [1]. Charge transport simulations of the double peak velocity-field characteristics showed a higher tuning frequency range due to the second peak. However, the conventional device structure used in this study did not include a hot electron injector which is known to reduce phase noise, improve temperature stability, and improve efficiency [2], [3]. Additionally, the energy balance simulation method provides a more accurate analysis than the drift-diffusion method used in the simulation. This work builds on previous work to understand the double Gunn effect using a more advanced simulation method and device architecture.

In this study, the second region of differential mobility in a Gunn diode with hot electron injection is varied, and the effect on the frequency and power output is presented. Section II discusses the device structure and the mobility models used in energy balance simulations. The implementation of the double Gunn effect into the standard temperature-dependent mobility models for GaAs is discussed in section III. Results are presented in section IV, and a conclusion is provided in section V.

2. Gunn diode with hot electron injection

Figure 1 presents the structure of a Gunn diode with a graded gap hot electron injector. The device contains highly doped contact layers on each end of the device and a transit region between them. A hot electron injector and a doping spike are placed between the cathode contact layer and the transit region.

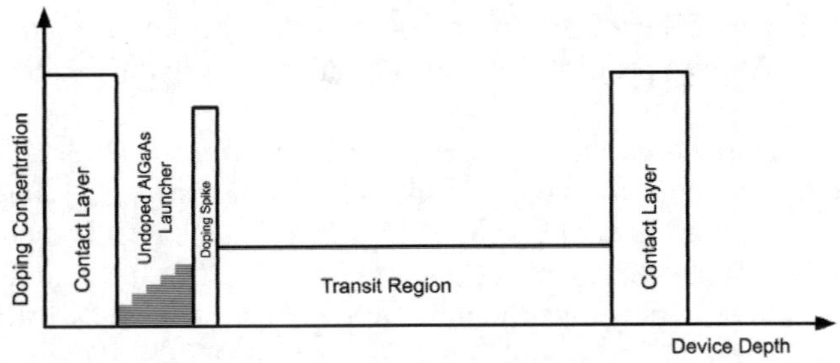

Figure 1. The structure of the Gunn diode with hot electron injection.

2.1. Graded gap injector

The addition of the hot electron injector has been shown to significantly decrease the phase noise and increase the temperature stability of the Gunn diode, which are both necessary for automotive radar applications [4]. In the conventional diode, carriers must gain energy by accelerating over a distance known as the dead zone. However, by passing through the injector, electron energy increases beyond the conduction band offset caused by the AlGaAs-GaAs junction. This effectively removes the dead zone in the transit region since the hot carriers leaving the injector contain enough energy to be scattered into the L valley [2], [3]. The turn-on voltage is reduced, and the efficiency is also increased by the presence of the injector.

Greater aluminum content in the AlGaAs compound increases the bandgap. Typically, mole fractions near 0.3 are used. It is common practice to increase the aluminum content in several discrete steps gradually. This grading of the launcher allows a particular voltage profile and response to increasing bias. The launcher is designed to be wide enough to resist low-energy tunneling but narrow enough to limit resistance. The doping spike following the launcher is required to achieve the desired electrical field in the transit region and avoid the development of a depletion region in the transit region [5], [6] [7] [8].

Table 1. The device lengths and doping concentrations used for Gunn Diodes with Hot Electron Injection.

Distance (μm)	Transit Region Concentration (cm^{-3})
0.7	3.5×10^{16}
1.1	2×10^{16}
1.65	1.1×10^{16}

The transit region lengths and doping concentrations that were used to observe the double Gunn effect are listed in table 1. The device architectures used in this study were originally used in commercially developed Gunn diodes. The concentration of the doping spike is 1×10^{18} cm^{-3}. Both contact layers have a doping concentration of 5×10^{18} cm^{-3}. The aluminum content in the graded gap injector is varied from .17 to 0.32.

2.2. *Physical models*

An energy balance simulation was used to provide a realistic depiction of the charge transport process in the device and, ultimately, the device's performance. By coupling the energy conservation equation with the charge continuity equation and Possoins equations, a more accurate depiction of electron movement is formulated that considers the nonlocal factors impacting the carrier's mobility. This consideration for carrier energy is necessary for this study since the double Gunn effect is inherently an energy-dependent process. A more primitive model could not capture the dead zone in the Gunn diode, for example.

Standard models for doping concentration-dependent low field mobility [9], electron saturation velocity [10], field-dependent mobility, negative differential mobility [11], and Shockley-Read-Hall recombination [12] were used for the simulation. The double Gunn effect is then implemented in these standard models to simulate the performance. These standard models have already shown satisfactory agreement with the commercially developed Gunn diode device [13]. These models, however, do not describe the double Gunn effect, which has only been shown in EMC simulation. Since EMC simulation data for the double Gunn effect is sparse and temperature-dependent, double Gunn mobility characteristics are defined for each doping concentration based on the abovementioned models and existing EMC simulation data. This process is detailed in the following section. The electrical characteristics depicting the double Gunn effect are only applied to the transit region, while the other regions maintained the standard mobility models.

3. Temperature-dependent mobility derivation

The behavior of the energy balance solver is sensitive to changes in the models due to the coupling of the additional equation. Therefore, the method used to apply double peak velocity characteristics to standard temperature-dependent mobility models must be constructed accordingly. The method shown in figure 2 is developed to realize the temperature-dependent mobility model for the double peak velocity characteristics. The dotted lines represent the use of a previous step to accomplish a later step. The process involves extracting standard mobility-field characteristics and temperature-mobility characteristics to form a temperature-field relationship. Then, the double Gunn mobility-field characteristics are used to define double Gunn mobility-temperature characteristics.

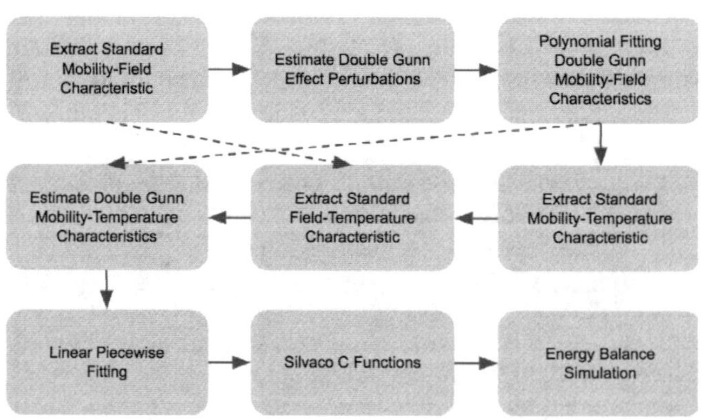

Figure 2. The process used to implement the Double Gunn effect into standard temperature-dependent mobility models in an energy-balance solver.

3.1. *Double peak mobility-field characteristics*

In order to define the field-dependent mobility characteristic of the standard models, I-V characteristic curves were extracted from simulations of homogenous GaAs. The GaAs sample must be doped at the necessary concentration since the mobility characteristic is concentration-dependent. Using the concentration and average electric field, a mobility-field characteristic is derived from the I-V characteristic. Next, several intensities of the double Gunn effect are gradually introduced to these characteristics by adding the velocity peak. Figure 3 shows the perturbations of the second velocity peak. It was necessary to study several intensities of the double Gunn effect to understand the phenomena. At higher compensation ratios and lower temperatures, the second-peak-to-valley ratio in GaAs is higher, indicating the strong presence of the effect. An approximation of this relation to temperature and compensation ratio is provided in table 2. The table summarizes data gathered using Monte Carlo calculations on GaAs with doping concentration 1x 10^{-17} cm^{-3} [14]. Polynomial functions define mobility as a function of the electric field for each perturbation.

3.2. *Standard mobility-temperature characteristic*

To calculate the temperature-mobility function for the standard models, a homogeneous GaAs sample is simulated again. However, instead of the I-V characteristic, this method uses the distributions (field, concentration, velocity, temperature) which result from a single bias. The sample must be doped at the same concentration as the one used for the mobility-field characteristic. Then, the homogeneous device can be discretized in one dimension. The carrier mobility and carrier temperature values at each cell in the one-dimensional mesh are used to form a mobility-temperature relationship. The field, concentration, and velocity distributions are used to solve for the mobility at each point in the discretized homogeneous sample. A very fine mesh must be used to obtain an accurate relationship. A large bias must be used so that the carrier temperature in the sample will have an effective electric field that spans the same range as the mobility-field characteristics. The standard temperature-dependent mobility model resulting from this process is presented in perturbation 0 (p0) of figure 4.

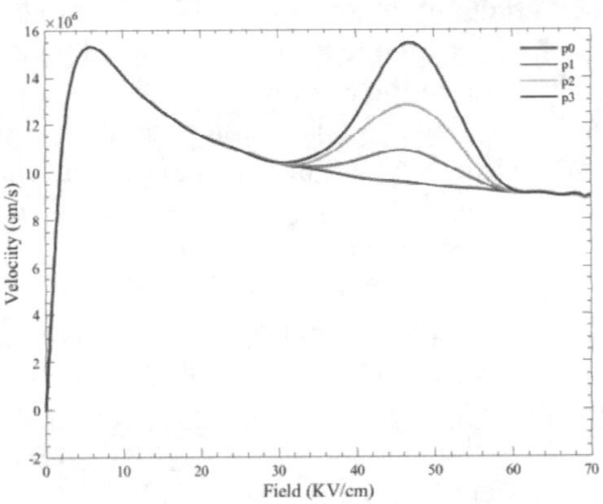

Figure 3. The velocity-field characteristics of typical GaAs with several intensities of the double Gunn effect.

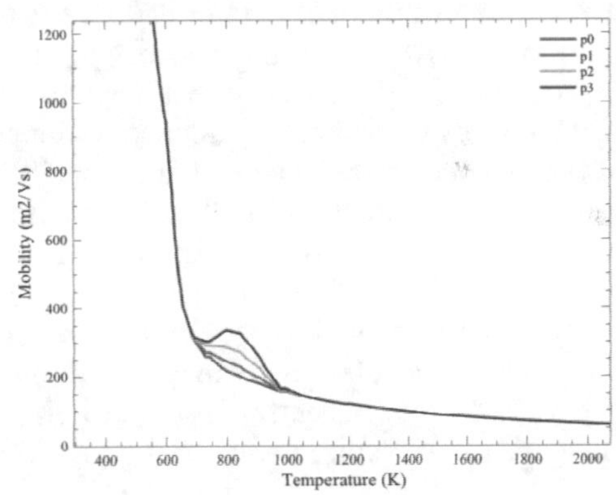

Figure 4. The mobility-temperature characteristics of typical GaAs with several intensities of the double Gunn effect.

Table 2. Temperatures and compensation ratios for various intensities of the double Gunn effect in GaAs with doping concentration 1x 10^{-17} cm^{-3} [14].

Perturbation #	Perturbation (%)	Compensation Ratio at 77K	Compensation Ratio at 160K	Compensation Ratio at 300K
0	0	--	0	0
1	22.5	0.15	0.2	0.6
2	50	0.4	0.4	--
3	100	0.8	0.7	--

3.3. Double Gunn mobility-temperature characteristics

The temperature-dependent mobility profiles for the double Gunn effects are presented in figure 4. The increase in the mobility for larger perturbations is due to the same process that causes the second velocity peak in figure 3. The characteristics in perturbations 1-3 are derived using the standard temperature-dependent and the perturbed field-dependent mobility characteristics. First, a standard temperature-equivalent electric field characteristic is solved using the standard mobility-field polynomial and mobility-temperature characteristic, though the roots must be handled appropriately. Using this temperature-field characteristic that represents the standard models, each mobility-field characteristic representing a given intensity of the double Gunn effect is used to solve for a temperature-mobility characteristic. Linear piecewise functions are constructed to describe these characteristics. These functions are then inserted in the energy balance simulation using C functions. Since the standard models have been compared to measurement data previously, the modified models should provide an accurate depiction of the double Gunn diode performance.

4. Results

A stepped bias profile is used to observe the performance of each device. A constant bias is applied for a 0.5 ns period, and the bias is incremented by 0.2 V over a 0.5 ns period. Seven discrete biases between 1.8 V and 3V are simulated. The frequency of the Gunn oscillations at each of the discrete biases at each bias is extracted from fast Fourier transforms. Stable oscillations were considered by ignoring the first oscillation period due to its transiency. The frequency and power achieved under bias variation are shown for each device in figure 5. The oscillation frequency increases with the intensity of the double Gunn effect in all devices. The frequency increases are also consistent across biases in almost all cases. In the device with a transit length of 0.7µm, the frequency gradually increases from 86 GHz to 98 GHz from perturbations 0 to 3. For the device with a transit length of 1.1µm, the frequency gradually increases from 57 GHz to 61 GHz for perturbations 0 to 2, and then perturbation 3 jumps up in frequency above 70 GHz. This large increase in frequency for perturbation 3, however, is accompanied by a drastic decrease in power. In the device with a transit length of 1.65µm, the frequency changes significantly with bias for all perturbations, unlike the other devices. However, the double Gunn effect still leads to larger frequencies for lower biases. At larger biases, the double Gunn effect causes a decrease in frequency, which increases the frequency range of the bias response. Overall, the linearity of the tuned frequency is not worsened by the presence of the effect. The frequency and power results in figure 5 show that a proportional decrease in power accompanies an increase in frequency. This power-frequency consistency occurs consistently across most biases in all devices.

The range of oscillation frequencies and average power for the bias response is plotted with respect to the intensity of the effect in figure 6. For devices with transit lengths of 0.7µm and 1.65µm, the increasing intensity of the double Gunn effect gradually causes a larger range of oscillation frequencies. However, this correlation between the intensity of the effect and the frequency range is weak, unlike that presented in the case of the conventional Gunn diode [1]. This is expected from the complex device geometry and the more precise energy balance simulation method used in this simulation. For the device with a transit length of 1.1µm, small intensities of the double Gunn effect don't increase the range of oscillation frequencies. However, perturbation 3 achieves a large frequency range accompanied by low average power. Furthermore, figure 6 shows that a power-bandwidth consistency persists in addition to the power-frequency relationship mentioned previously. As the perturbation changes, increases or decreases in the frequency range for a perturbation often lead to directly proportional changes in average power.

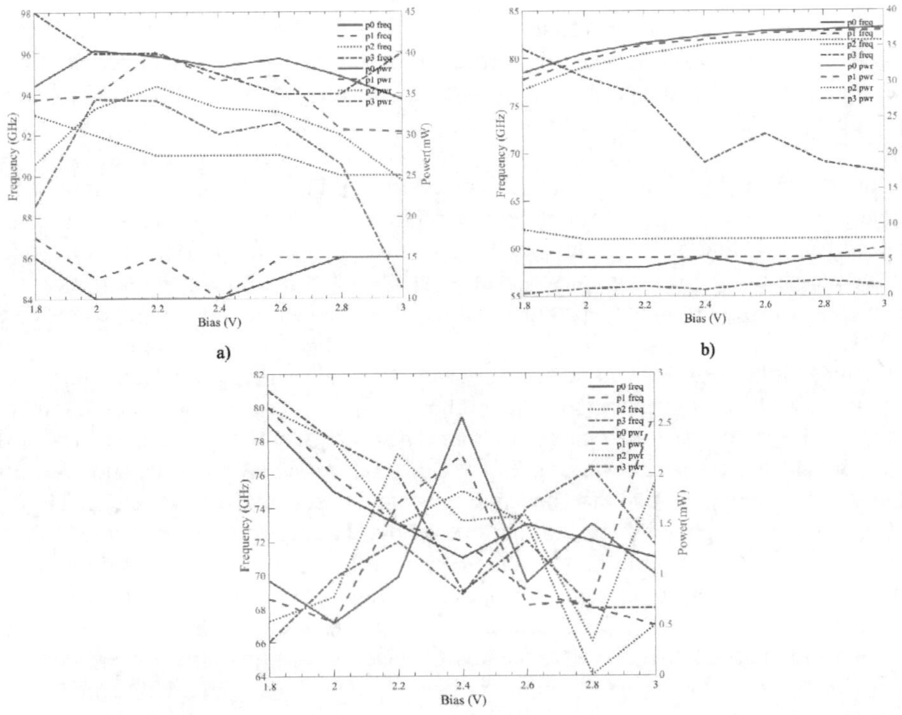

Figure 5. Oscillation frequency as a function of bias for different intensities of the double Gunn effect in device with transit lengths of a) 0.7μm, b) 1.1μm, and c) 1.65um.

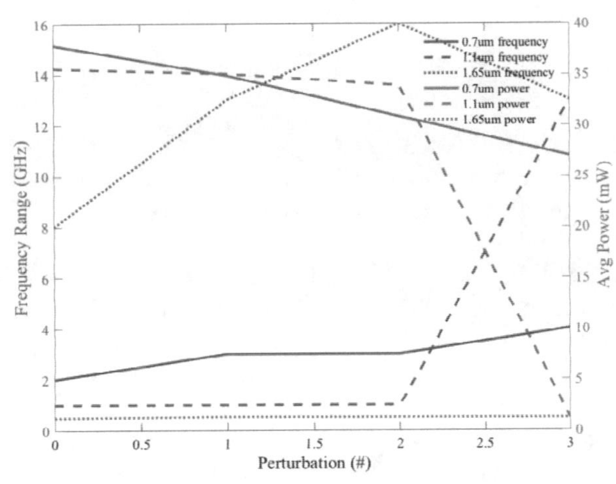

Figure 6. Frequency range and average power achieved in each device for different intensities of the double Gunn effect.

Figure 7 presents the power-frequency results for increasing intensities of the double Gunn effect in each device. Devices with transit region lengths of 0.7μm and 1.1μm are presented in figure 7a, and figure 7b shows a closer look at the response for the device with a 1.65μm transit length. The left y-axis represents the average of the power-frequency product at each bias. The right y-axis represents the product of the average power and the frequency range. This can be referred to as the average-power-bandwidth product. Figure 7a shows that the average power-frequency product decreases with the size of the device. It also shows that the product of the average power and the frequency range decrease with the size of the device. In the device with a transit length of 1.1μm, these values are almost constant for different perturbations 0-2, while perturbation 3 shows a noticeable decrease. This follows the deviation for perturbation 3 described in figures 5 and 6. For devices with transit lengths 0.7μm and 1.65μm, the power frequency and power frequency range remain relatively constant for increasing intensities of the double Gunn effect. In the shorter device, the average power-frequency product decreases as perturbation increases, but the average-power-bandwidth product has the opposite trend. In the longer device, the average power-frequency product hardly changes while the average-power-bandwidth product doubles.

The significance of these results can be summarized in the following manner. In the smallest device, the frequency increase shown in figure 5a is outweighed by the power loss. However, the increase in the frequency range in figure 6 provides a more favorable power tradeoff for all intensities of the double Gunn effect. In the device with a transit length of 1.1μm, the frequency increase shown in figure 5b is met by a similar decrease in power so that their product remains constant. However, the largest intensity of the effect causes a large power decrease that outweighs the large frequency increase in figure 5b. Therefore, the characteristics in perturbation 3 are less useful for efficient high-frequency oscillations. Still, perturbation 3 provides a significant increase in the tunability of oscillations. The largest device shows lower output power than the other devices, but it achieves much higher frequency ranges. In this device, the double Gunn effect provides a frequency increase for a comparable power decrease. Like the smallest device, the double Gunn effect improves the average-power-bandwith product, making the power decrease well worth the increase in tunability.

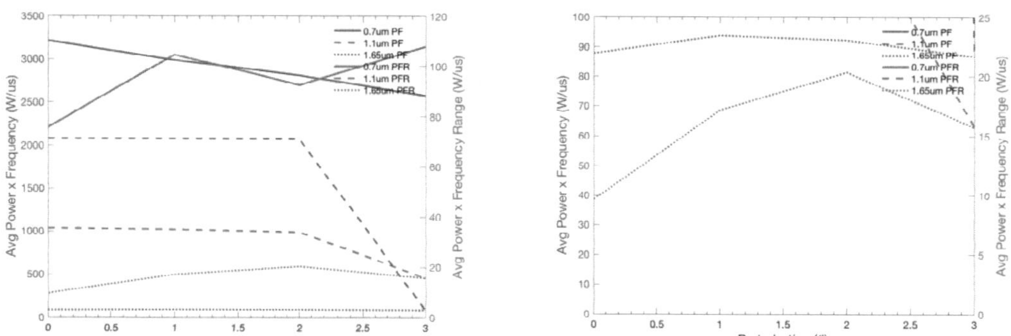

(a) (b)

Figure 7. The product of the average power-frequency product and product of average power and frequency range for devices with transit lengths a) 0.7μm and 1.1μm, and b) 1.65μm. "PF" refers to the average of the product between power and frequency. "PFR" refers to the product of the average power and the frequency range for the biases considered.

It should be mentioned that the three simulated devices are considerably different, although their structures are similar. When changes in the relatively complex device architecture are considered along with the sensitivity of the coupled energy balance model, unique domain dynamics and device performance can be expected. This is evident in the device with a transit length of 1.1μm which achieves oscillation frequencies lower than those in the larger device. While the 1.65μm device does have a larger frequency, its output power is much lower. Although the cases in this study span a large region of interest, further continuity between them is required to observe more consistencies and improve characterization. The introduction of additional cases, however, quickly increases the difficulty of the study due to the computational costs of the energy conservation solver. Despite minor differences between device performance, the responses to the double Gunn effect appear consistent for all devices.

From the results of this study, it is clear that the double Gunn effect leads to larger oscillation frequencies and tuning ranges, which occur at the cost of lower output power.

5. Conclusion

This work has presented the power-frequency and power-bandwidth performance of a step-graded Gunn diode operating under the double Gunn effect. A commercially developed device architecture and standard models were selected for investigation. Standard models for GaAs temperature-dependent mobility were methodically modified to include the double Gunn effect. Four intensities of the effect were tested for three devices. A consistent increase in oscillation frequency and tuning frequency range is observed. The power output remains comparable across the range of biases for each device. These simulations indicate that compensated GaAs could provide increasingly larger oscillation frequencies and tuning bandwidths as temperature decreases while maintaining the advantages of hot electron injection. More test cases and further analysis of the domain propagation are required to provide a complete description of the device's performance. Additionally, the impact of X valley injection into the transit region should be investigated for its impact on frequency tuning.

References

[1] L. Butts and S. El-Ghazaly, "The Effect of Double-Peak Velocity Characteristics on Gunn Oscillations," in SoutheastCon 2022, Mar. 2022, pp. 487–490.
[2] S. Neylon et al., "State-of-the-art performance millimetre wave gallium arsenide Gunn diodes using ballistically hot electron injectors," in IEEE MTT-S International Microwave Symposium Digest, Jun. 1989, pp. 519–522 vol.1.
[3] N. R. Couch, M. J. Kelly, H. Spooner, and T. M. Kerr, "Hot electron injection in millimetre wave Gunn diodes," Solid-State Electron., vol. 32, no. 12, pp. 1685–1688, Dec. 1989.
[4] N. Priestley, K. Newsome, I. Dale, and P. Norton, "A Gunn diode based surface mount 77 GHz oscillator for automotive applications," in 2002 IEEE MTT-S International Microwave Symposium Digest (Cat. No.02CH37278), Jun. 2002, vol. 3, pp. 1863–1866 vol.3.
[5] S. A. M. Akhbar and D. S. Ong, "Analysis of notch-δ-doped GaAs-based Gunn diodes," J. Phys. Appl. Phys., vol. 55, no. 37, p. 375103, Jul. 2022.
[6] N. Farrington, M. Carr, and M. Missous, "Rapid evaluation of doping-spike carrier concentration levels in millimetre-wave GaAs Gunn diodes with hot-electron injection," 2010.
[7] N. Farrington, M. Carr, J. Sly, and M. Missous, "Development of an on-wafer test for rapid evaluation of doping spike carrier concentration levels in commercially manufactured GaAs Gunn diodes for automotive radar applications," Jan. 2008.
[8] C. Lee and U. Ravaioli, "Monte Carlo comparison of heterojunction cathode Gunn oscillators," Electron. Lett., vol. 26, pp. 425–427, Mar. 1990.
[9] D. M. Caughey and R. E. Thomas, "Carrier mobilities in silicon empirically related to doping and field," Proc. IEEE, vol. 55, no. 12, pp. 2192–2193, Dec. 1967.
[10] R. N. Hall, "Electron-Hole Recombination in Germanium," Phys. Rev., vol. 87, no. 2, pp. 387–387, Jul. 1952.
[11] J. J. Barnes, R. J. Lomax, and G. I. Haddad, "Finite-element simulation of GaAs MESFET's with lateral doping profiles and submicron gates," IEEE Trans. Electron Devices, vol. 23, no. 9, pp. 1042–1048, Sep. 1976.
[12] W. Shockley and W. T. Read, "Statistics of the Recombinations of Holes and Electrons," Phys. Rev., vol. 87, no. 5, pp. 835–842, Sep. 1952.
[13] F. Amir, N. Farrington, T. Tauqeer, and M. Missous, "Physical Modelling of a Step-Graded AlGaAs/GaAs Gunn Diode and Investigation of Hot Electron Injector Performance," in 2008 International Conference on Advanced Semiconductor Devices and Microsystems, Oct. 2008, pp. 51–54.
[14] E. Y. Wu and B. H. Yu, "Monte Carlo calculation of temperature dependence of the transport properties in compensated GaAs," J. Appl. Phys., vol. 70, no. 5, pp. 2719–2724, Sep. 1991, doi: 10.1063/1.349386.

AUTHOR INDEX

Institute of Physics
Dirac House, Temple Back
Bristol BS1 6BE UK

ISSN: 1742-6588
ISBN 978-1-7138-7703-5

9 781713 877035

CURRAN ASSOCIATES INC.
proceedings.com